SELECTIVE FIRE

UZI

Semi-Auto Modification Manual

Desert Publications
El Dorado, AR 71730 U. S. A

Selective Fire
UZI
Semi-Auto Modification Manual

505 East Fifth Street
El Dorado, AR 71730 U.S.A.
870-862-3811
www.deltapress.com

ISBN 978-0-87947-004
10 9 8 7 6 5 4
Printed in U.S.A.

The Publisher Desert Publications is a division of
The DELTA GROUP, Ltd.
Direct all inquiries & orders to the above address.

Neither the author nor the publisher assumes any responsibility for the use or misuse of the information contained in this book. This material was compiled for educational and entertainment purposes and one should not construe that any other purpose is suggested.

Warning!!

The Publisher (Desert Publications) produces this book for informational purposes and under no circumstances advises, encourages or approves of use of this material in any manner.

Table Of Contents

Foreword

The popularity of select fire weapons has always been high with a segment of the shooting fraternity. By provision of Federal law, it is possible for a civilian, if in agreement with state and municipal laws, to lawfully make or purchase a select fire weapon. Of course the weapon must be transferable and the tax ($200) be paid.

Once over this hurdle, the prospective buyer of a machine gun, because of limitations of the Gun Control Act of 1968, must confine his purchase to a weapon that is not of foreign manufacture. Imported machine guns are limited to law enforcement agencies and to Class 3 dealers as sales samples.

The only way to legally circumvent this is to purchase a semi-automatic version of the desired weapon and, by way of a BATF approved Form 1, convert it yourself. You may, also, purchase such a weapon that has been converted by a Class 2 manufacturer. In the latter case, the transfer takes place the same as for an American made machine gun.

For various reasons some imported weapons lend themselves to being converted to select fire easier than others. Naturally, if the conversion is too simple, the BATF will usually take some action to hinder the ease of conversion and, hoepfully, curtail the traffic of contraband weaponry.

The purpose of this book is to show the individual how they can legally make a select fire conversion of one weapon in particular – Israel's famous Uzi. Both the author and the publisher discourage anyone from using the knowledge contained herein in any unlawful manner.

Class 2 manufacturers will find this information of particular interest. While the methods described will produce a near perfect weapon which approaches factory original select fire models, only home workshop tools are required.

Historical Background

Military rifles and sidearms have been with us almost since gunpowder was first introduced to Europe, and machine guns were a fact well before the turn of the century. Submachine guns, however, are uniquely a Twentieth Century weapon, like airplanes and submarines.

For those unfamiliar with the basic definition, submachine guns are simply automatic weapons which are chambered for a pistol round, but are somewhat larger than a pistol. Most have a shoulder stock (either collapsible or fixed) and can also be fired from the shoulder in the semi-auto mode. Although they can be fired in the full auto mode with one hand, they are meant to be held with two hands in the underarm assault position when fired in this manner, no matter what Hollywood might have you believe.

The first SMG to see active service in any quantity was the Italian Villar Perosa, a rather odd looking device with two barrels and dual magazines. Although originally designed for use in aircraft, the Villar Perosa was quickly adopted for infantry use.

The German MP18I first appeared in 1918. Its *Maschinen Pistole* designation is misleading, since it actually looked very much like an ordinary autoloading rifle, complete with shoulder stock. With troops stalemated in the trenches for months on end, something besides the traditional Mauser, SMLE and Springfield battle rifles was needed for the close in trench fighting. The MP18I was the German answer to the problem while the Americans favored the Winchester M97 shotgun. However, SMG development had begun in the U.S. and Brig. Gen. John T. Thompson was pushing hard to get some samples of his SMG to the front lines for battlefield testing before the war ended. Legend has it that the first European-bound shipment of Thompsons arrived in New York on November 11, 1918, the same day the armistice was signed.

Although the war ended before the effectiveness of the MP18I could be fully analyzed, the Allies were sufficiently impressed with it that the Treaty of Versailles specifically forbid Germany to manufacture SMG's or issue them to her 100,000 man army. This prohibition did little good, as SMG development continued underground until the 1930's, by which time it was no secret that Germany was rearming at a fast pace, and not just with defensive thoughts in mind.

Submachine gun development in the U.S. ground to a standstill following the end of the war. Indeed, Thompson was never able to sell a great number of his guns, and wasn't even a major stockholder in any of the firms that manufactured it. Total production figures for the Thompson in all models during the 20 years spanning 1919 to 1938 barely exceeded 10,000 units. Thompson died in

1940, never to know that in less than five years "Tommy gun" production would exceed the 2,000,000 mark.

Design work continued in Europe, however, and by the 1930's numerous countries had adopted subguns as standard issue. Some outstanding designs were produced by Beretta and Schmeisser, as well as other manufacturers. The Spanish Civil War proved an excellent testing ground for SMG's, showing the weapons' overall superiority for house to house and forest fighting.

Neither the U.S. nor Britain had an issue SMG when war broke out in 1939, although a small number of Thompsons had been used by the U.S. Marine Corps in China and Nicaragua in the 1920's. The British ordered a large quantity of Thompsons after the debacle at Dunkirk and went to work on an SMG design of their own. The result was the famous Sten, in 9mm Parabellum. Although the standard British service pistol cartridge was the .38-200, which was a rimmed revolver cartridge, they adopted the 9mm simply because its characteristics were already well known and its adoption would enable them to use large stores of captured ammunition.

America's entry into the subgun sweepstakes was the M3 "Grease gun", which looked very much like its namesake. The M3 and the Sten both met the major requirements for a wartime SMG admirably – they were rugged, reliable and cheap to produce. The Russian equivalent was the PPSh 41, an extremely crude yet efficient weapon which looked far more like a rifle than the SMG's we know today. The German MP38 with folding stock was already clearly recognizable as an SMG. It was redesigned into the MP40, which was quite a bit cheaper to produce. All of these weapons were of blowback design, which is practically universal among the breed.

The end of World War II saw a cessation of hostilities in most of Europe. While the Soviets quickly communized Eastern Europe, British troops found themselves locked in battle with a new enemy – Jews in Palestine. One member of the Haganah who joined the British Army to steal weapons was a 17 year old youth named Israel Galili (spelled Uziel Gal by some sources). A native of the Galilee area, he had a mechanical aptitude and was called upon to modify and repair the various weapons which fell into Haganah hands. This gave him the opportunity to study their designs and ascertain their good and bad points. The experience gained while so employed was to serve him well in laters years, enabling him to become one of the foremost firearms designers in the world.

When the British pulled out of Palestine in 1948 and the state of Israel was born, the Israeli arms industry no longer had to remain underground. Israeli Military Industries was formed, with Israel Galili joining them almost immediately.

The value of the submachine gun in modern warfare was well known to IMI, and they charged Galili with designing a native weapon for use by Israeli troops. He borrowed elements from the Czech M23 and 25 and came up with the Uzi, which was adopted for service in 1951.

Probably no other nation on Earth is as war-minded as Israel, simply as a matter or survival. Surrounded by hostile Arab neighbors, the Israelis grow up with the idea they probably (not maybe) will be called on to lay down their lives for their country. They want weapons that will continue to function in the harsh desert climate, where a firefight and sandstorm may batter at them simultaneously. The Uzi is such a weapon, as is the new Galil assault rifle which is based on the famous Kalashnikov AK-47 which functions *anywhere* and under any conditions.

The Uzi quickly earned a reputation for ruggedness and reliability in the numerous wars which Israel has been perpetually fighting ever since she first came into being. Such a reputation hardly goes unnoticed in military circles, and numerous other Free World countries have adopted the Uzi, which is manufactured by IMI and also under license by Fabrique Nationale in Belgium, one of the most respected and successful arms manufacturers of all time.

Although not a standard issue SMG with U.S. forces, the Uzi is still employed by some government agencies, notably the Secret Service. Who doesn't recall the photos of the Secret Service agent brandishing an Uzi just seconds after President Reagan was shot in 1981?

The Uzi is an outstanding example of what is known as second generation submachine guns, smaller and more compact than first generation guns like the Thompson and MP38. Its cyclic rate of 650 rounds per minute is well above the "ideal" subgun rate of 450 r.p.m., but few if any of the SMG's in service achieve that ideal. The Uzi's rate is still slow enough that even someone who has never fired an automatic weapon before can hold his bursts down to 3 shots each while really experienced subgunners can achieve single shots even when the selector lever is set for full auto.

Although there are newer designs available, the Uzi will probably be with us for a long, long time due to its outstanding track record. Its human engineering is better than that found on most SMG's and its grip safety is an added plus. Many people complain about the grip safety of the Colt Government Model, but the Uzi's is designed in such a way as to be no problem at all.

Although hardly capable of benchrest shooting, the Uzi's accuracy is more than adequate for the job for which it was designed. SMG's are intended for close quarters use on full auto and, with their pistol cartridges, have a maximum effective range on semi-auto of roughly 200 yards. Disabling or lethal hits beyond that range are possible, but hardly probable.

Since the 1970's, there has been a growing market for semi-auto only subgun look alikes. A number of shooters want the look and feel of an automatic weapon without having to pay the $200 transfer tax to legally own one. Due to certain provisions of the Gun Control Act of 1968, it would be impossible for the average American citizen to own a full auto Uzi even if they were willing to pay the tax. Recognizing the existence of this market, IMI introduced their semi-auto only version in 1980. It was an instant commercial success and was back ordered with the distributor for some time until the overburdened supply finally caught up with the demand.

Semi-auto Uzis are instantly recognizable by their longer barrels which are necessary to make them legal. Dummy 10.2" barrels are available to give them the same appearance as the full auto version. Of course, the weapons can't be fired with the dummy barrels in place. Fitting an actual 10.2" barrel to a semi-auto Uzi would be in violation of the National Firearms Act of 1934, which restricts the minimum legal barrel length for rifles, carbines and shotguns. However, the shorter replacement barrel can be installed on a legally converted gun since it is already registered with the BATF as a machine gun.

Early full auto Uzis were originally made with a wooden buttstock, and some even had wooden pistol stocks instead of the grooved plastic ones we are familiar with today. Metal folding stocks were later added as an option. Today both the full auto and semi-auto Uzis are normally supplied with the folding stock. However, wooden buttstocks identical to the originals are available as accessories and are highly desirable for more comfortable shooting.

Although Americans can't legally buy an Uzi submachine gun made in Belgium or Israel, they can convert the semi-auto version to full auto only or selective fire with prior BATF approval. They can either have this done by a firm specializing in such conversions, or they can do it themselves. Legal home workshop conversions were being done within months of the semi-auto's appearance on the market. Full details on the selective fire conversion are described and illustrated in the following pages.

Of special interest to automatic weapons buffs is the new Mini Uzi. Although scaled down in size, it fires the same 9mm Parabellum round as its larger brother. With a cyclic rate of 1250 rounds per minute, recoil compensation is a must. It has compensator cuts in the top of the barrel, just behind the muzzle. The Mini Uzi is aimed at the law enforcement and security markets rather than the military and should enjoy the same wide acceptance that the full size subgun has achieved.

Given the economic and political turmoil that is present in the world today and that may well grow worse before the situation improves, a growing number of people are acquiring legal automatic weapons as a hedge against a total breakdown of society as we know it. Their detractors look upon them as doomsayers and laugh at their preparations, yet one need only look at what happened in much of Europe following the end of World War II. Literally millions of homeless, starving people were uprooted, killed or enslaved as the Communists seized Poland, Czechoslovakia, Hungary, Yugoslavia, Albania, Bulgaria, Romania and a large part of Germany. The people in these countries had been disarmed during or before the German occupation and had no way of resisting the Red tide.

There is an old saying that history repeats itself. Actually, it would be more correct to say that people just repeat the same mistakes their forefathers made in previous generations, so the possibility exists for the same thing to happen here.

It's also a well known fact that an enemy finds it extremely difficult if not impossible to invade a country where the citizenry is heavily armed. Switzerland is an excellent example. Its militia consists of every able-bodied male citizen between the ages of 20 and 60, and voters are required by law to wear sidearms (either pistol or sword) to the polls to show they are citizens in good standing. Switzerland hasn't been involved in a war for over 100 years. Although the natural defenses provided by the surrounding mountains are probably the main reason, an armed Swiss citizenry would still be a formidable deterrent to any would-be conqueror.

By contrast, England has very stiff gun control laws and it's for all practical purposes virtually impossible for the average citizen to own a handgun. Rifles and shotguns are also controlled. Yet when a German invasion appeared imminent during World War II, the phrase "Send a gun to defend a British home" was heard throughout America. The flow of American arms to Britain wasn't the only or even the main reason Hitler decided against invasion, of course, but it's a certainty he realized the enormity of the task should German troops have to face an armed British citizenry.

If Americans wish to preserve their freedom for the immediate and distant future, they must fight vigorously to protect the Constitution and all of our freedoms, including the right to legally own automatic weapons for the small percentage of citizens who want them. Should a social or economic collapse occur, or America be invaded by hostile forces from other lands, these people and their automatic weapons may well be in the vanguard of those who will be fighting to protect and preserve what is left of the American dream and way of life.

Mechanical Background

The Uzi semi-auto carbine, as sold in abundance in the United States, is made by Israeli Military Industries, which was the original manufacturer of the selective fire version. This "street legal" sub-gun look-alike was introduced to satisfy a growing U.S. market for such weapons. Since the Gun Control Act of 1968 precludes the importation of full automatic weapons (except for law enforcement use and as Class 3 dealer samples), a semi-auto version was a logical solution.

While Israel and her Major Israel Galili are praised for their genius concerning this weapon, the fact remains they merely copied the Czechoslovakian Zk 746 and Models 23 and 25 in which the use of a telescoping bolt was introduced.

Backed by American dollars and industrial know-how, Israel has maintained good quality control and production facilities. The unknowing payoff to us is availability of a very good weapon to convert to selective fire.

The Uzi semi-auto carbine probably displays more engineering effort to prevent ease of conversion than most such weapons. IMI made doubly sure that none of the critical parts used on the SMG model would fit the semi. The barrel length, to be U.S. legal, had to be at least 16" long and the chamber end was machined to a smaller diameter to prevent an original Uzi SMG barrel from fitting.

Next came the bolt and firing mechanism. The original Uzi SMG fired, typically, from an open bolt condition. The semi bolt was redesigned to fire from a closed position, using a separate firing pin and striker assembly. This was done to utilize the original sear. Since the entire stroke of the original bolt wasn't required, a large rubber buffer block was added in the rear of the receiver. Because a shorter bolt stroke was used, the cocking knob slot in the top cover was shortened.

To prevent an original SMG bolt from being substituted, a bar of steel was welded inside the right rear of the receiver and the semi bolt had a groove machined in it, allowing it to pass over the blocking bar. To further limit the semi-auto bolt's usefulness in a full auto conversion attempt, 1/2 of the bolt face rim was machined off. Since it would be chambering rounds with the firing pin retracted, it functioned satisfactorily. However, should an attempt be made to slam fire with the firing pin protruding, feeding of rounds from the magazine would be hampered by the pin.

Next, the designers turned their attention to the trigger mechanism and housing. In the original SMG the selector lever had three positions: safe, repeat and auto. Internally, the lever had a curved finger on the front that, in the repeat or semi-auto position, served to trip the disconnector. In full auto position this finger moved forward, clearing the disconnector. To prevent the lever from being able to move into the full auto position, a small block of steel was welded to the trigger housing.

Earlier Conversion Methods

Some of the first attempts at converting the semi-auto Uzi to full automatic were, simply, removing the finger from the selector lever. Since the disconnector was no longer tripped, the sear would remain down until the trigger was released. Had the rim of the bolt face not been machined off, full auto fire would have been achieved, although semi-auto would have been lost.

Various alterations to the bolt were tried, including restoring the rim of the bolt face by welding a piece of steel to it. This worked, with the first shot fired from a closed bolt position and the remaining burst from slam fire. The striker assembly was heavy enough and the spring strong enough to detonate the primer when allowed to follow the bolt forward.

The first commercially available SMG type bolt was what is referred to as the short stroke bolt. It featured a fixed firing pin within a full 360° bolt face rim. It used the original recoil spring and rubber buffer and had a groove machined in it to clear the blocking bar. Because of the shorter stroke, the cyclic rate was higher than that of the original Uzi SMG.

Shortly after the short stroke bolt, along came a similar bolt designed to operate full stroke with a new recoil spring/buffer assembly. Having a longer stroke, it required the length of the cocking knob slot to be increased. Both of these bolts were used in some of the first legal conversions by Class 2 manufacturers.

Unfortunately, they were apparently used in a number of illegal conversions, or at least made such conversions a reality to those so inclined. As a result, the BATF, in a liberal interpretation of the law that states that a machine gun, besides the obvious, can be any part or group of parts that can readily restore or convert a semi-auto weapon to full auto fire, ruled that both of these bolts were machine guns. The groundwork for this interpretation had, no doubt, been laid when the AR-15 auto sear decision went down the year before. For these bolts to be sold without a National Firearms Act transfer, the groove had to be welded up or never there in the first place.

This all evolved into the bolt design that is now available without transfer and tax requirements. This bolt nearly duplicates the original Uzi SMG bolt. It will not function unless the user either machines a groove in the side of the bolt or removes the blocking bar from the receiver. In either case a machine gun has been manufactured and the responsibility to have done so legally lies heavily upon the person doing it.

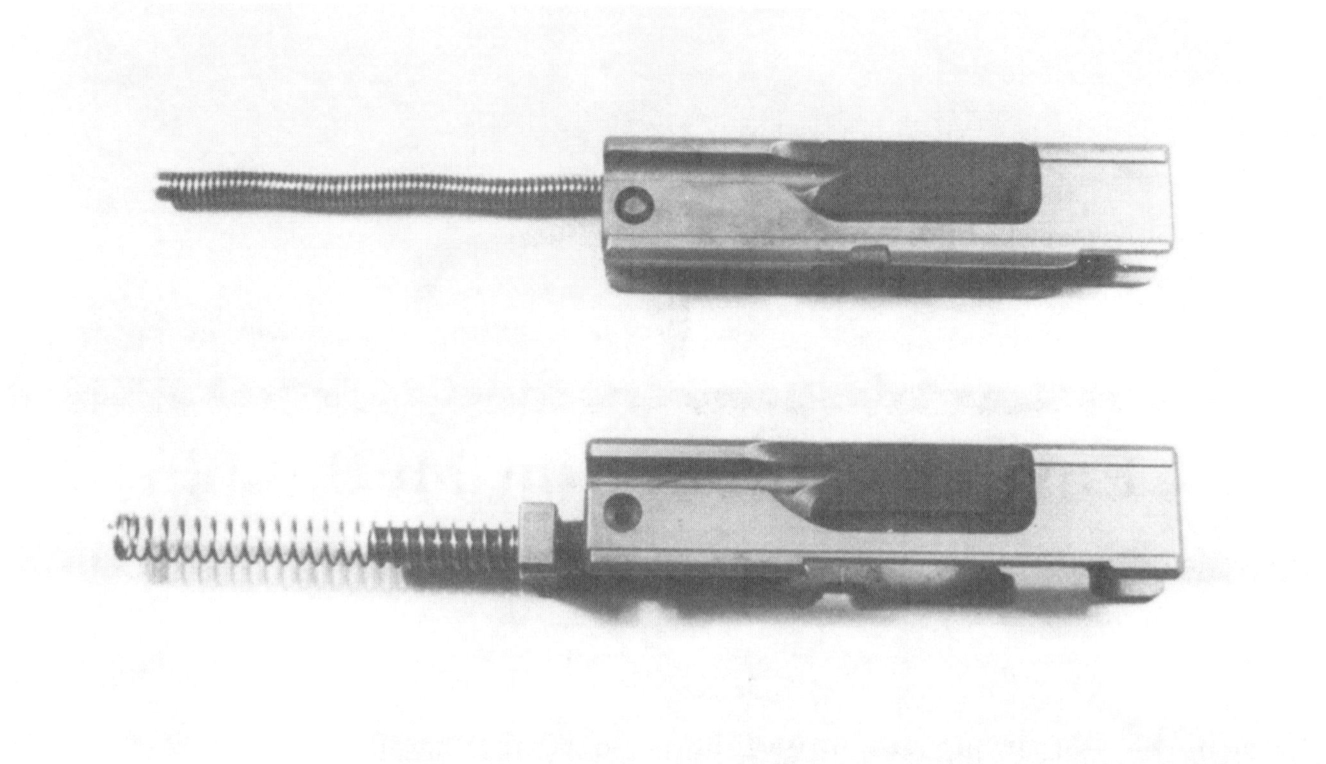

At bottom is an original Uzi semi-auto bolt & striker assembly. Missing from photo is the recoil spring. At top is an early replacement bolt that allowed open bolt firing. Now classified as a machine gun by the BATF, this bolt was used in early select fire conversions.

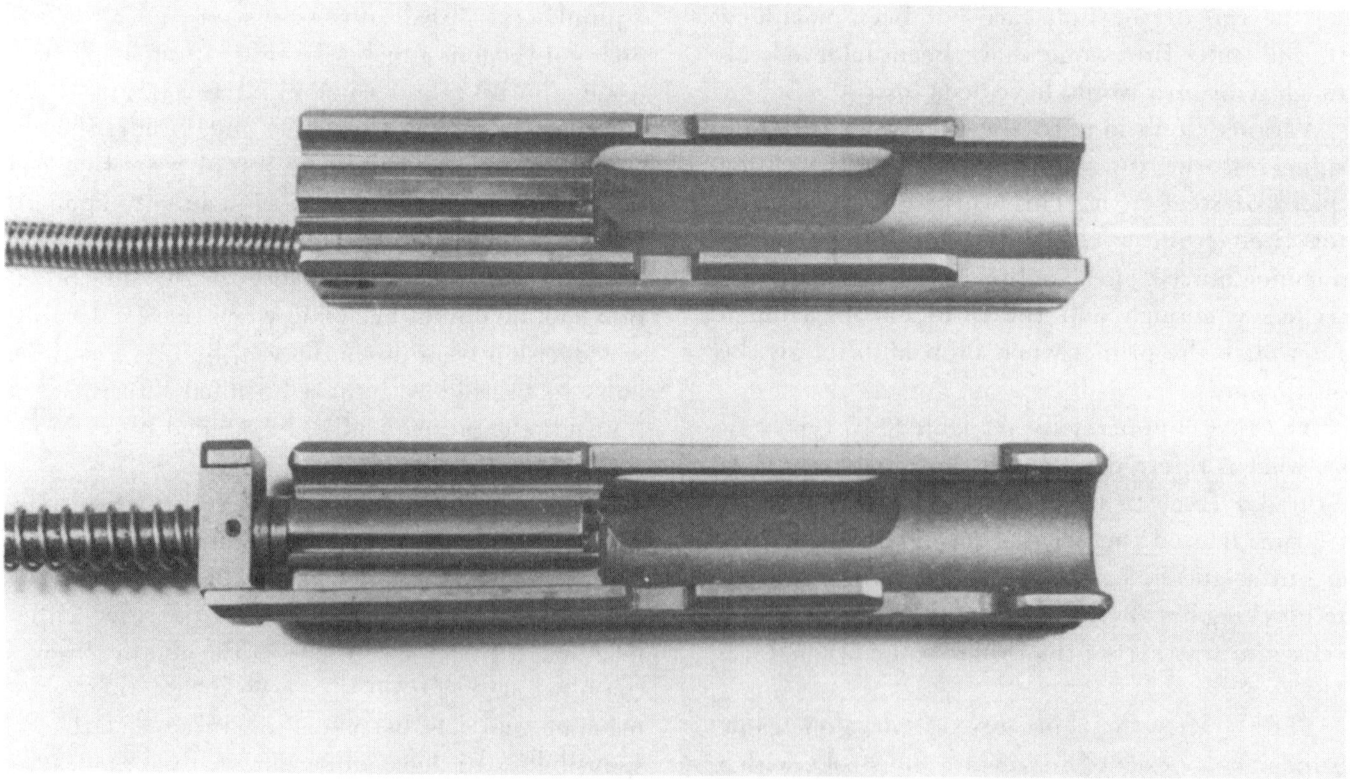

Bottom view of the same bolts.

Bolt face view of original bolt (right) and replacement bolt (left). Note how the rim of the bolt face (arrow) was machined off on the original.

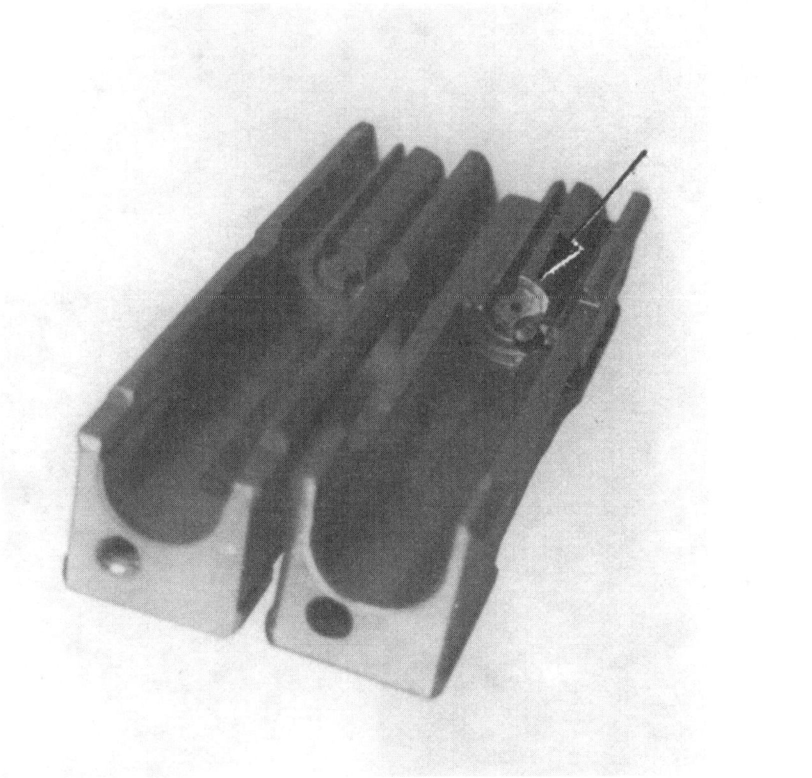

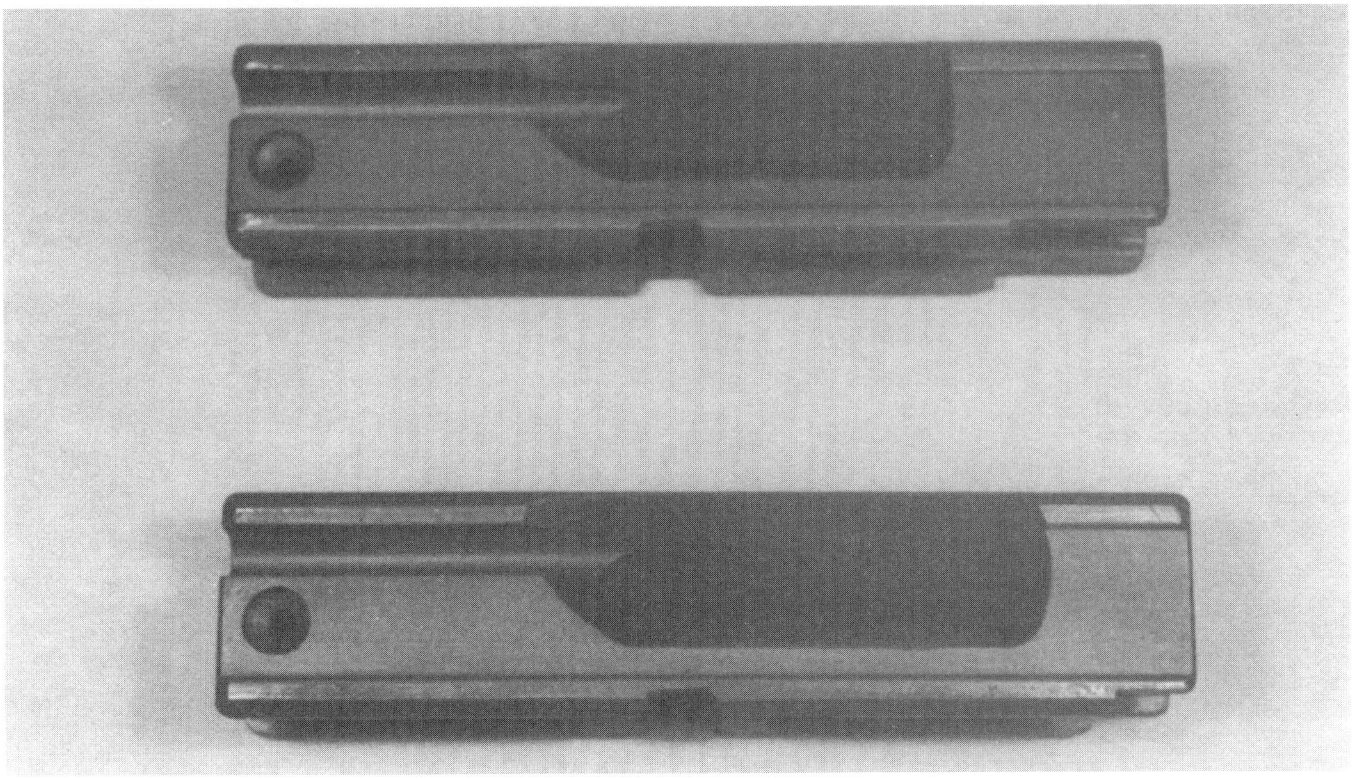

The original semi-auto bolt (top) compared to full stroke replacement bolt. The replacement bolt previously shown was referred to as the short stroke bolt while this one is referred to as the full stroke bolt. Like the short bolt, this one is also classified by the BATF as a machine gun!

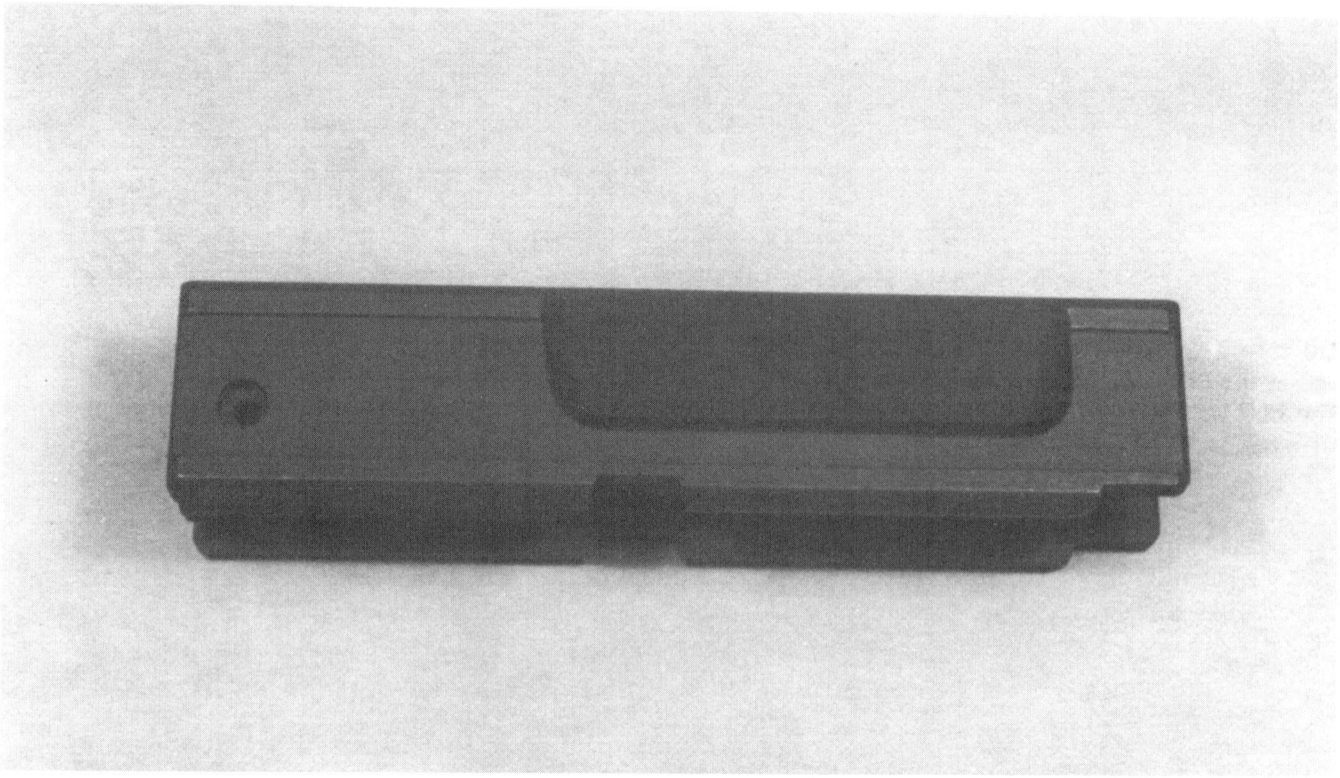

The bolt shown here is the only configuration that can be purchased without an NFA transfer and tax. Note that it is devoid of the receiver block clearance groove that the other bolts had.

Latest Conversion Method

With the availability of bolts that fire from the open position and various other parts, the most effort required in converting a semi-auto Uzi into a select fire submachine gun is removing the blocking bars from the receiver and trigger housing. The usual method has required a vertical milling machine to remove the bar from the trigger housing and to partially remove the receiver blocking bar.

With the method described herein, a milling machine is not required. To better understand this method, let's examine how IMI attaches the blocking bars. Mass production necessitates the blocking bars be made with locating tabs extending from the mating surfaces. These tabs locate into mating slots punched in the sheet metal receiver and trigger housing and are plug welded from the outside. Ground flush before any finish is applied, the welds are undetectable.

The trigger housing blocking bar has only one such tab while the receiver bar contains two, one

A new Uzi carbine before conversion to select fire.

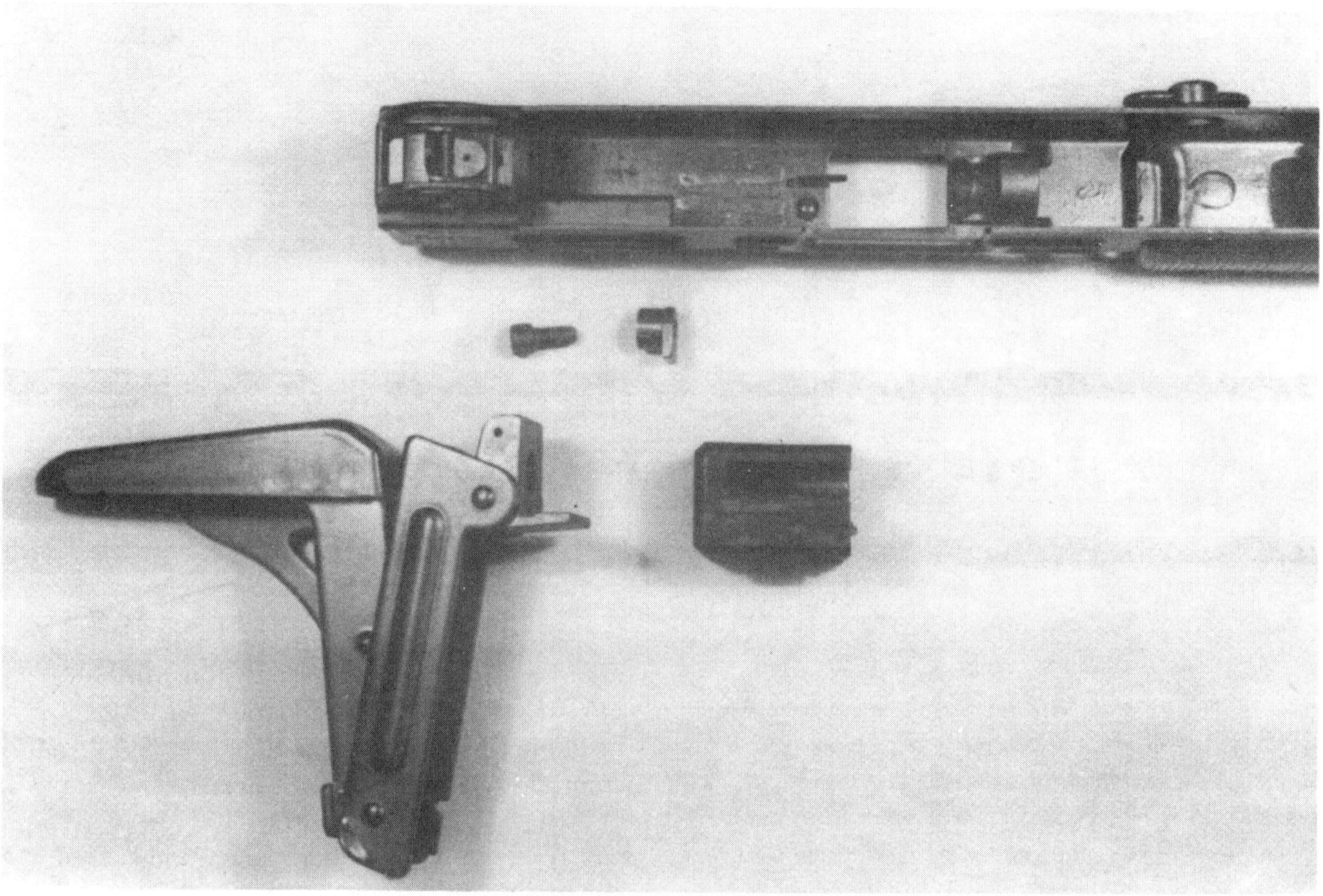

In order to remove the receiver block it is necessary to remove the folding stock and buffer block.

near each end. Each of the three tabs measures approximately 1/8" x 5/8". These tabs, because of the way they project from the blocking bars, are vulnerable to a shearing force. To remove them we will apply a twisting force that shears them flush with the sheet metal.

To apply such torque to the block in the trigger housing, a simple tool is made from a cheap 1/2" drive socket. The size of the socket should be as large as practical. The socket pictured in this book was 11/16", although any size in the neighborhood of 3/4" would do. This socket is altered by grinding a notch on each side, directly opposite each other. The size of these notches should fit over the blocking bar which is .320" wide and .156" high. Since the socket is very hard, the notches will have to be ground.

By using a long extension handle on the socket, torque can be applied to the block, which shears off very easily. In most cases a little clean-up with a Dremel tool grinder leaves the area smooth and flush. The bare metal can be protected by cold bluing if desired.

The same type of twisting force is used on the receiver blocking bar. However, it must be applied differently. Because that bar has two tabs, one at each end, we must cut it in two before twisting. While various ways could be utilized to cut it, the simplest is to drill it nearly in two as illustrated on page 16. Locate and center punch the starting point as shown in the drawing. This point is just behind the front tab and leaves enough of the remaining bar protruding from under the rear of the receiver to grab and twist.

It is advisable to start with a smaller drill bit, say 1/8" first, and increase in size until the drill just clears the sheet metal side of the receiver. By this time it should have already broken through the

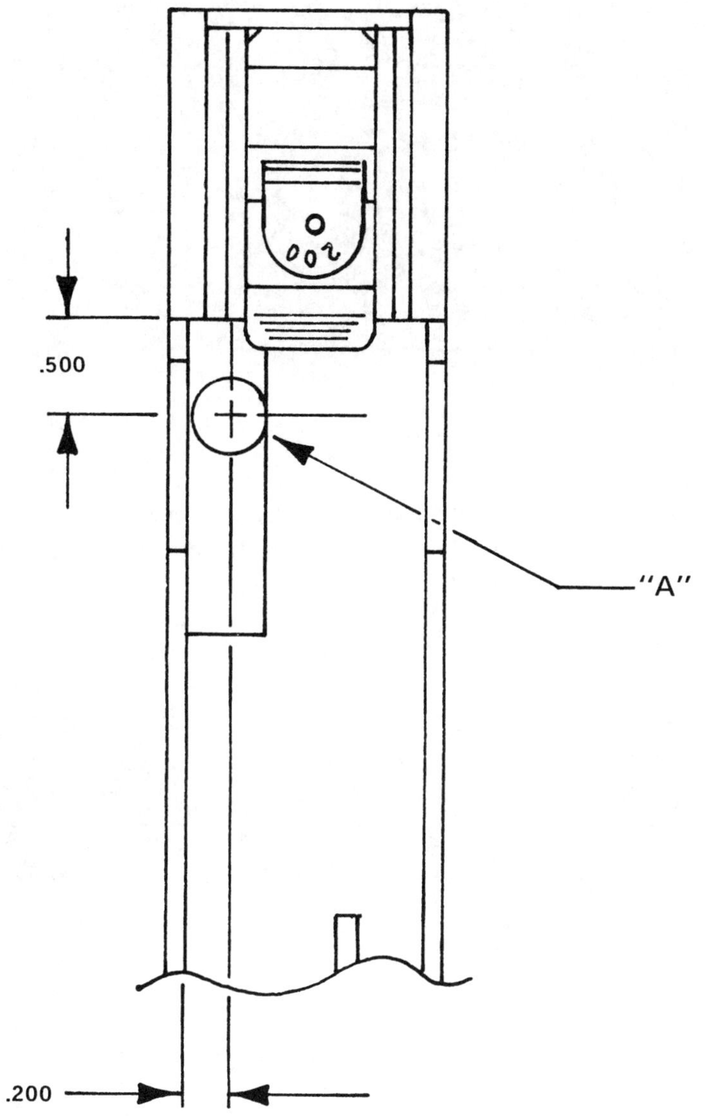

This drawing shows the location to drill through the receiver block. Start hole "A" with a smaller bit and increase the diameter of each successive hole to approximately .375" diameter. Use care not to let the drill bit touch the sheet metal receiver side plate.

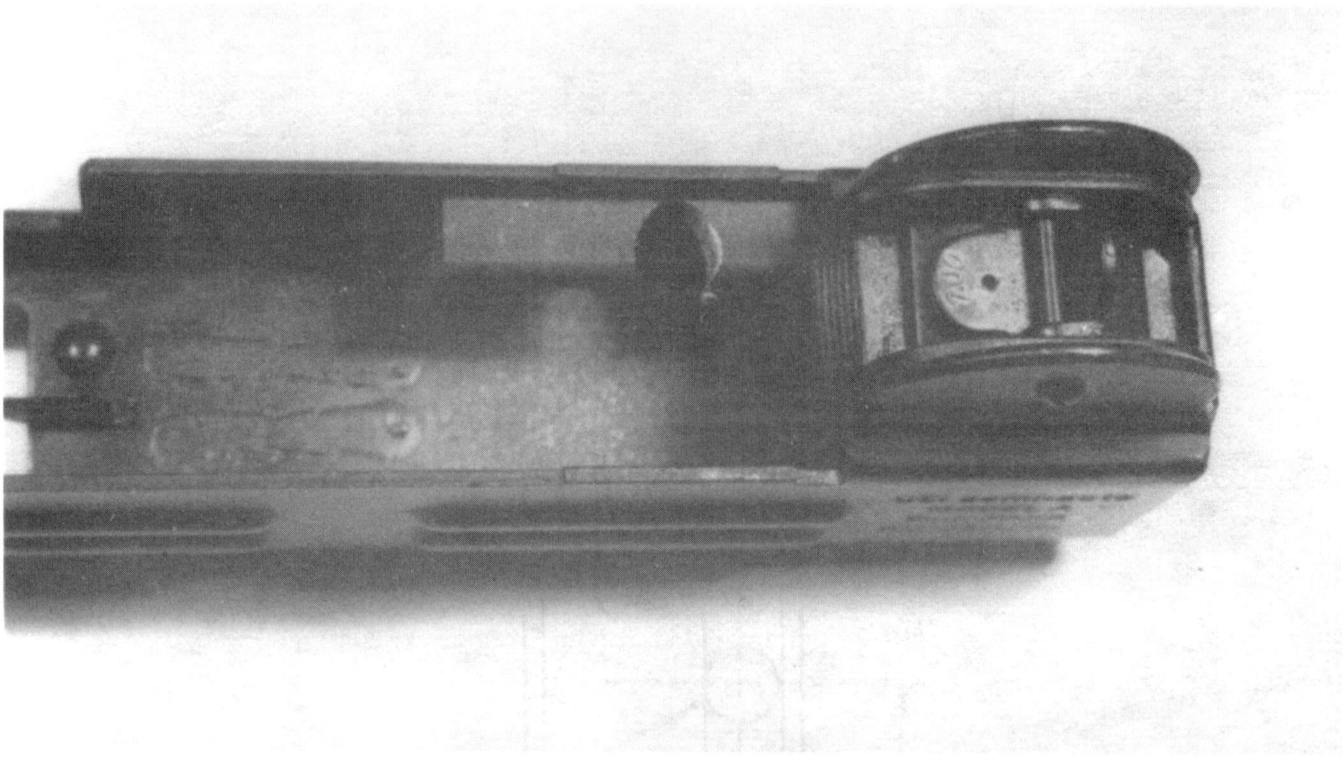

Your finished hole should look similar to this. The receiver block may not be quite severed. However, this will break through okay.

Grab the front piece of the block with an adjustable wrench as shown and twist up and down.

Two or three times should be sufficient to shear it free as shown here.

The remaining part of the block is removed the same way.

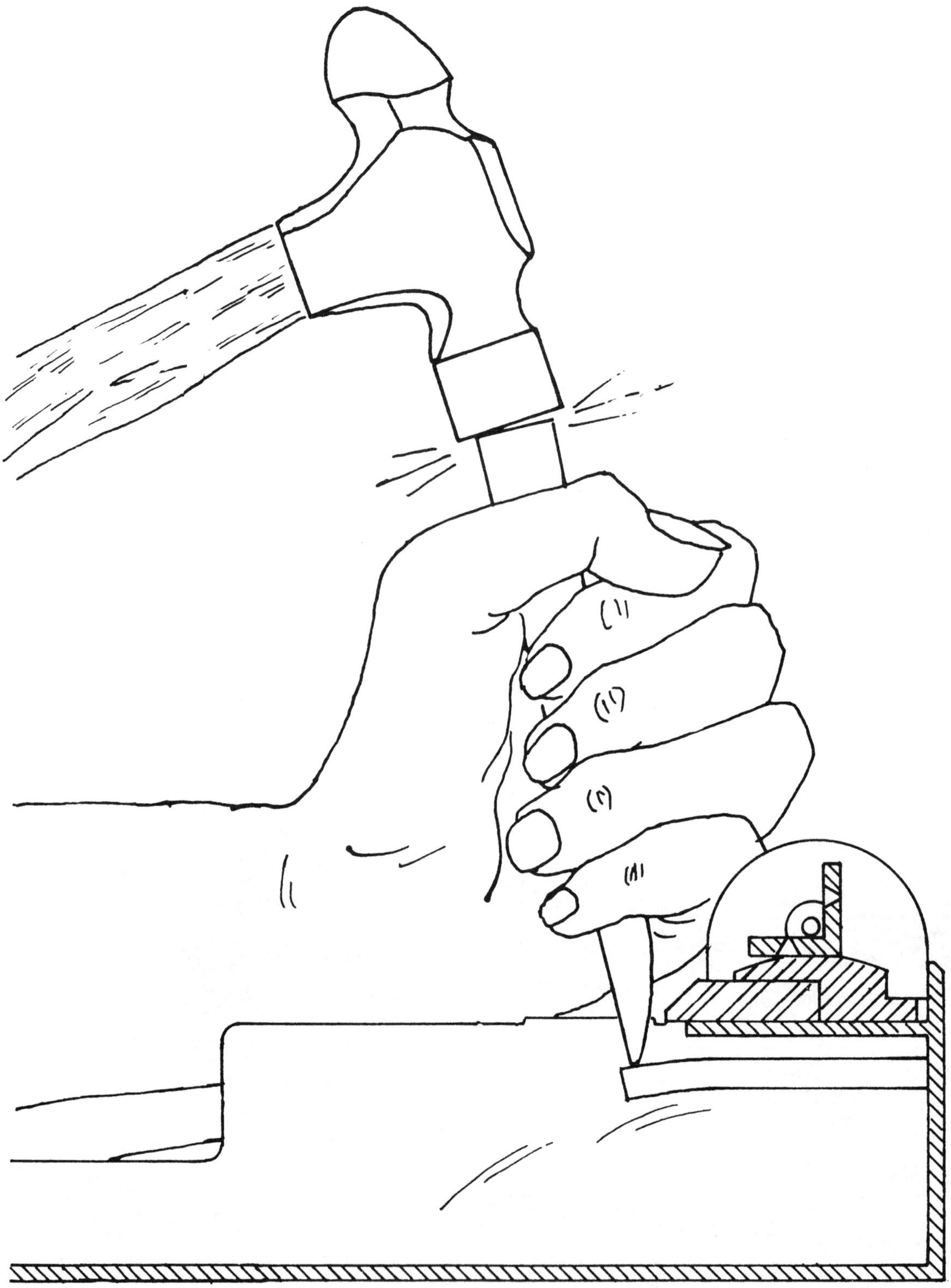

Since the travel of the wrench may be movelimited on the rear part of the block, a hammer and punch can be used to tap the block further down. Use a large, flat-bladed screwdriver to lift it up. 2 or 3 movements like this and the rear part will shear off flush with the receiver. A Dremel tool grinder can be used to smooth up the side where it can reach. If a smoothing is necessary on the rear area, a small flat file can be used to reach it. A touch of cold blue and you can hardly notice where the block was.

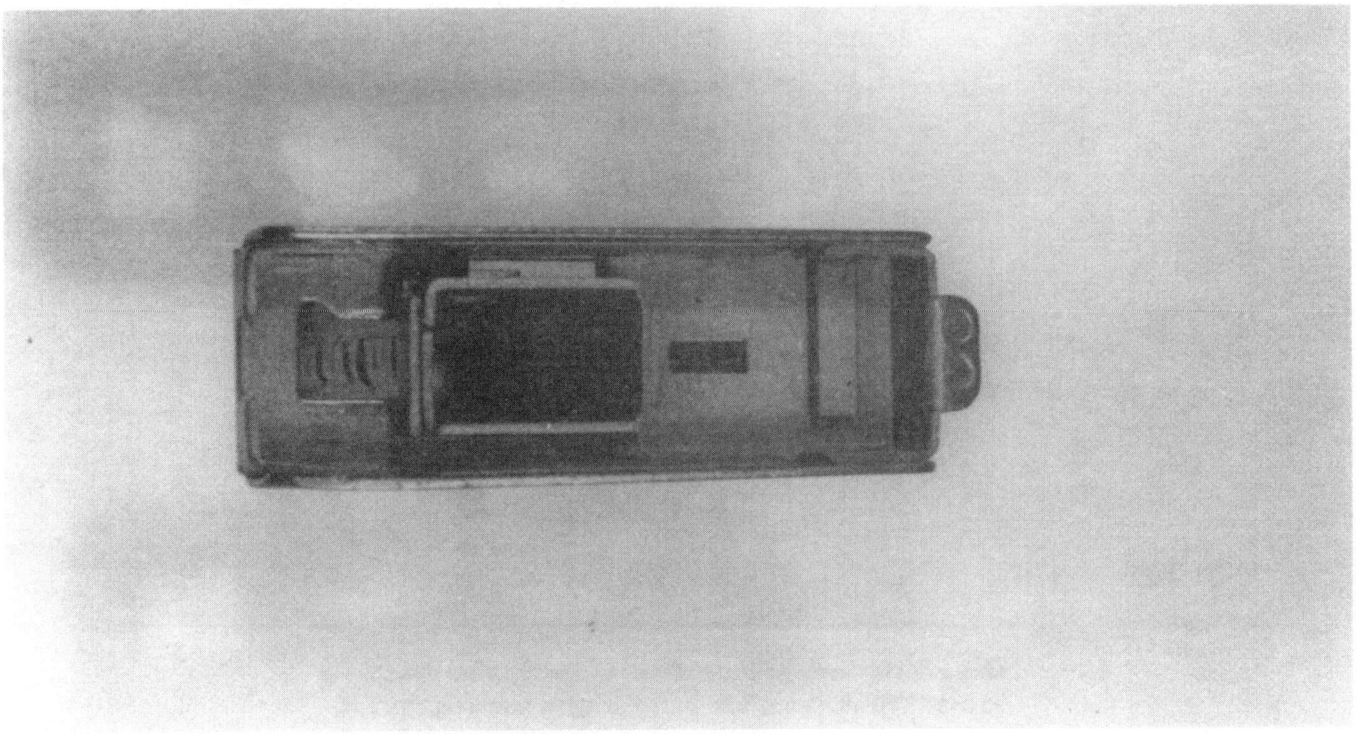

For a near to factory original SMG, the selector blocking bar in the trigger housing must be removed.

To remove this bar we must make a tool to apply torque to it. Obtain a cheap 3/4" socket in 1/2" drive. Use a metal cutting saw to cut a notch in each side.

This notch need only be deep and wide enough to fit over the blocking bar in the trigger housing.

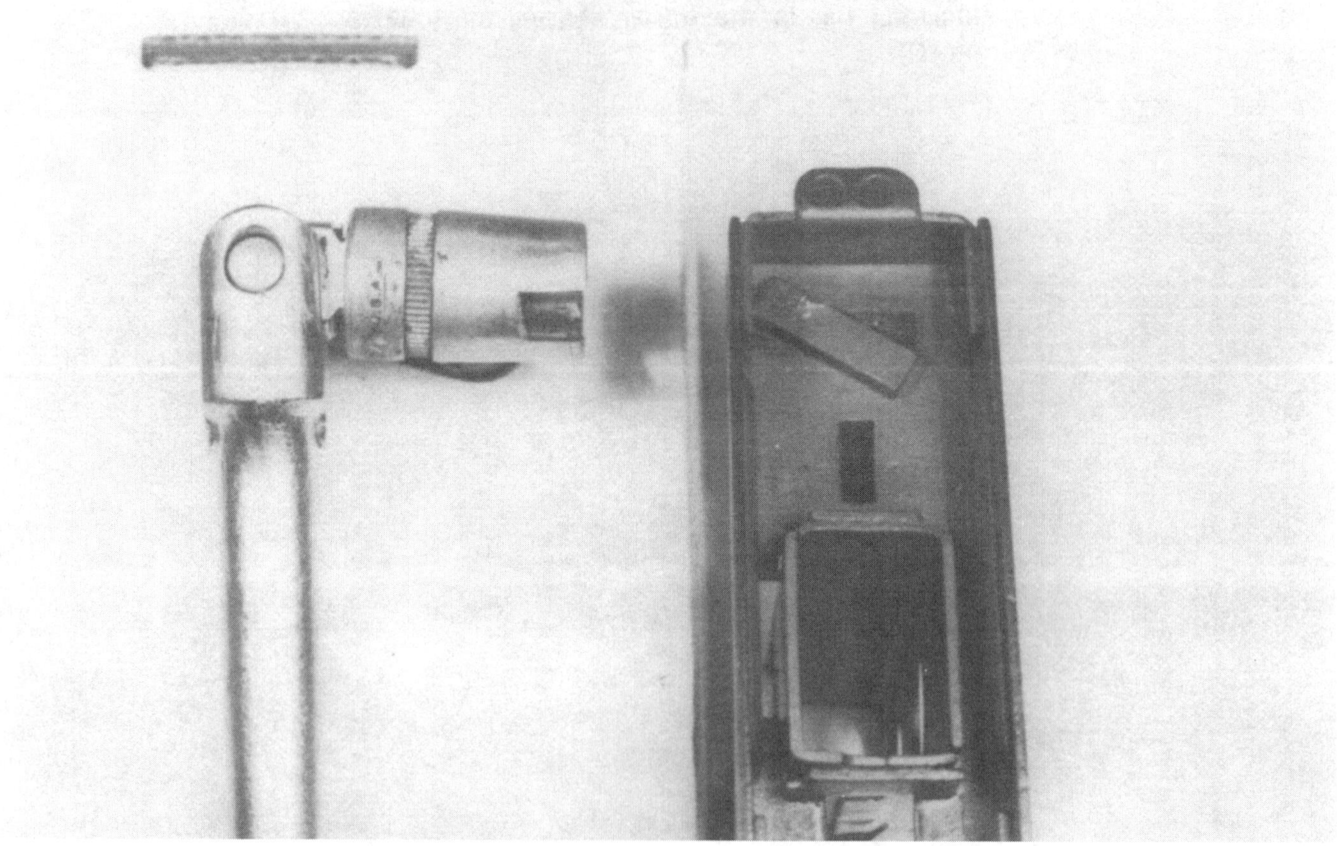

Using the altered socket on an extension handle, torque is applied to the blocking bar until it shears flush with the housing.

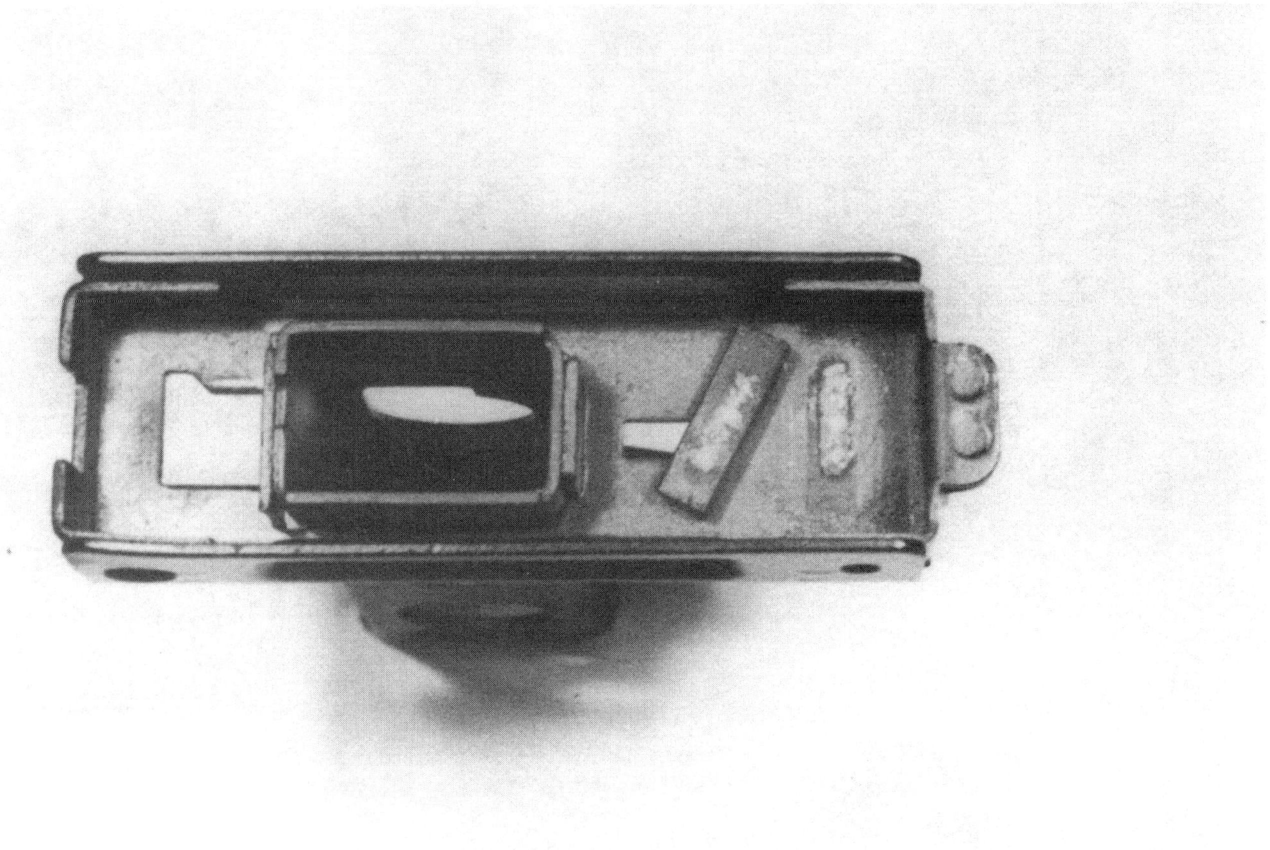

This photo shows how the block had a tongue that fit into a slot in the sheet metal. It was plug welded from below and ground flush. For a perfect job you can smooth the area with a Dremel tool grinder. Protect the bare metal with cold blue.

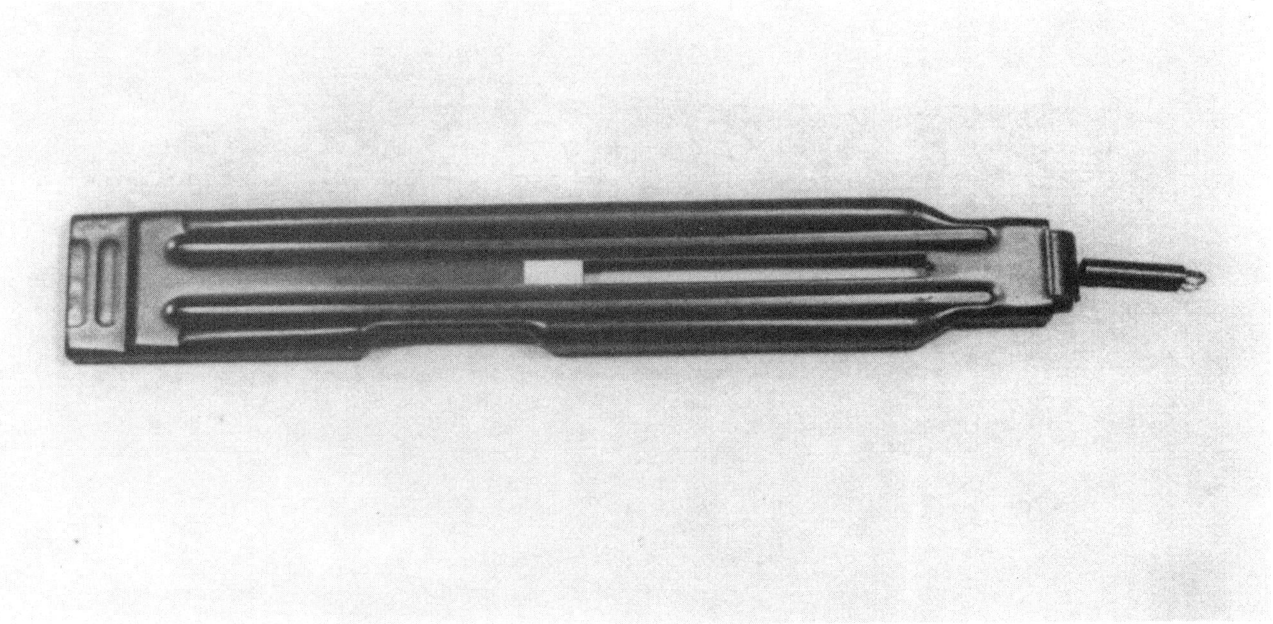

Since the new bolt we'll be using travels further to the rear, it is necessary to elongate the cocking knob slot by 5/8". A piece of white tape on the cover plate illustrates the area to be removed.

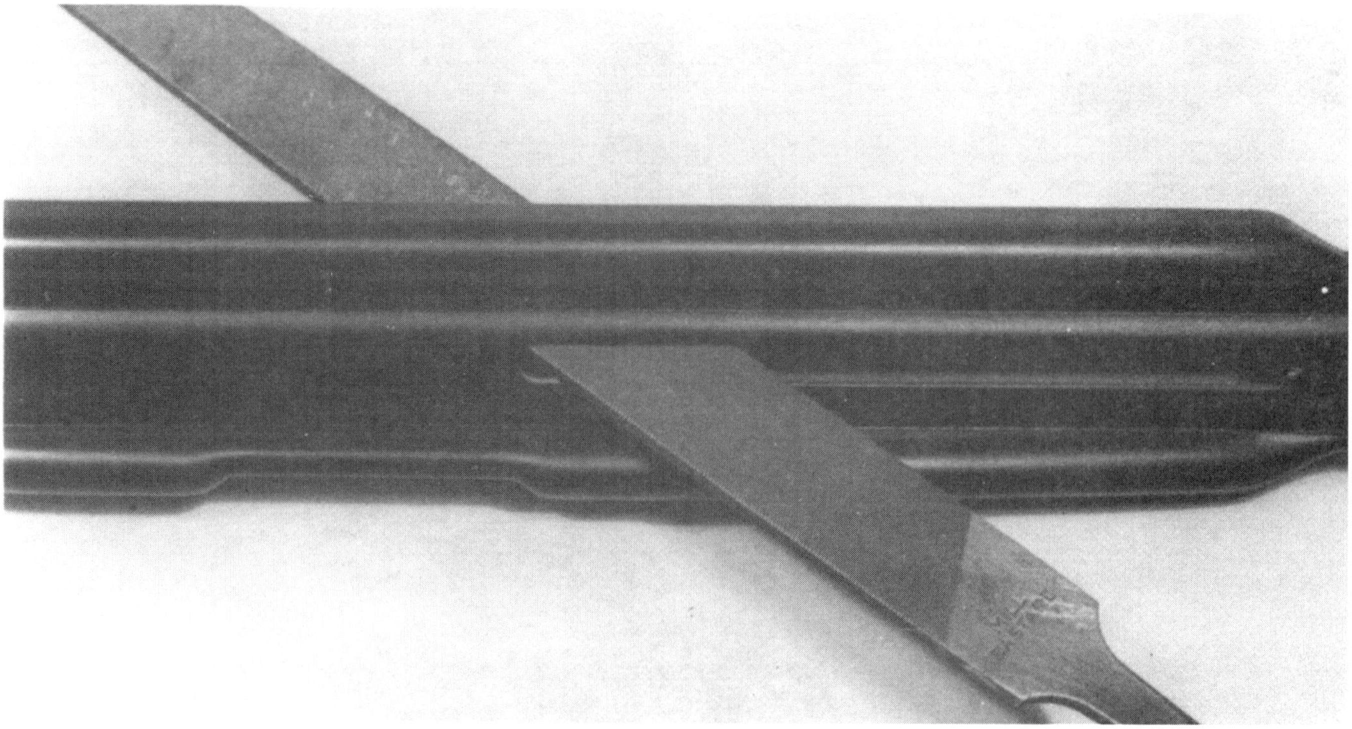

Although a milling machine would make the job easier, a flat file can be used to elongate the slot.

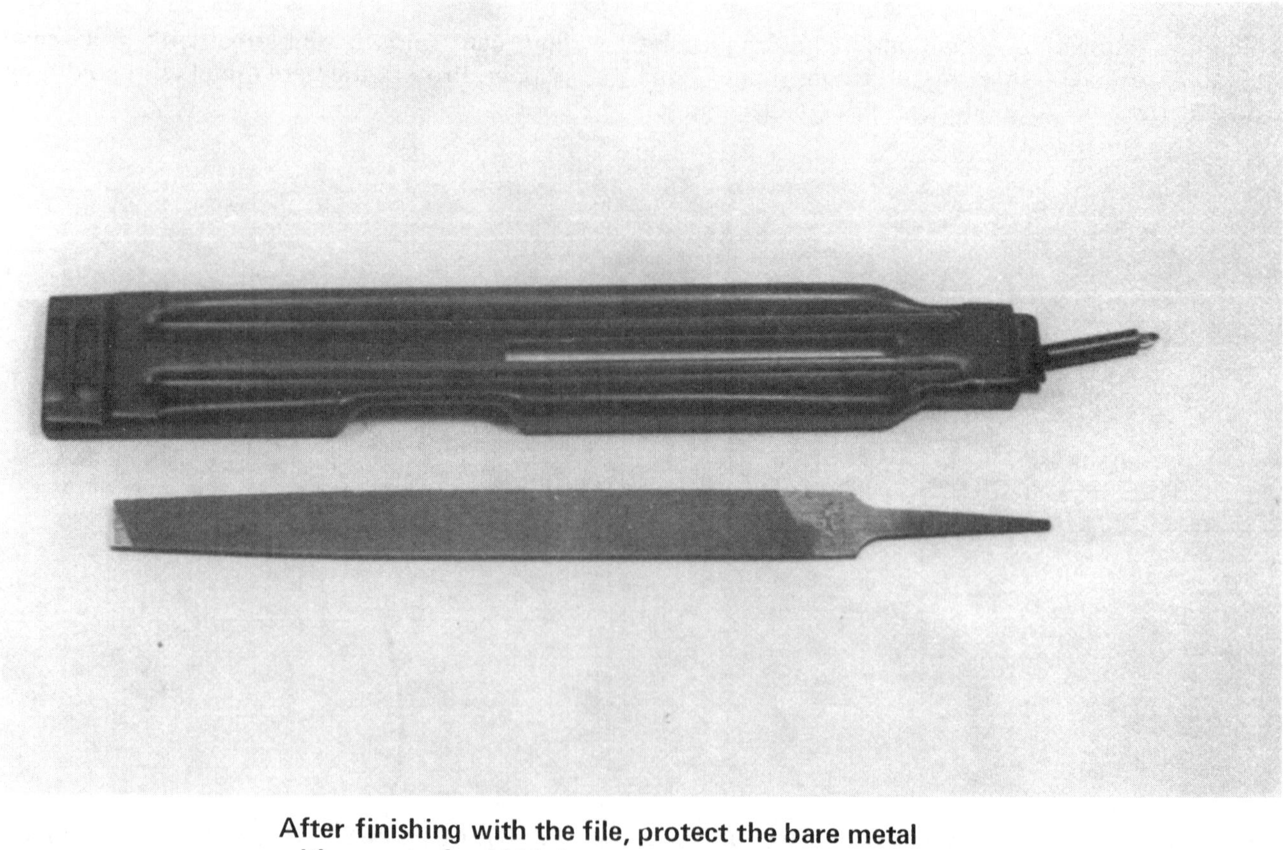

After finishing with the file, protect the bare metal with a coat of cold bluing.

The original selector lever (lower) has a safety extension but contains only 2 detents. A replacement lever (upper) is available that has all 3 detents but is devoid of the safety extension.

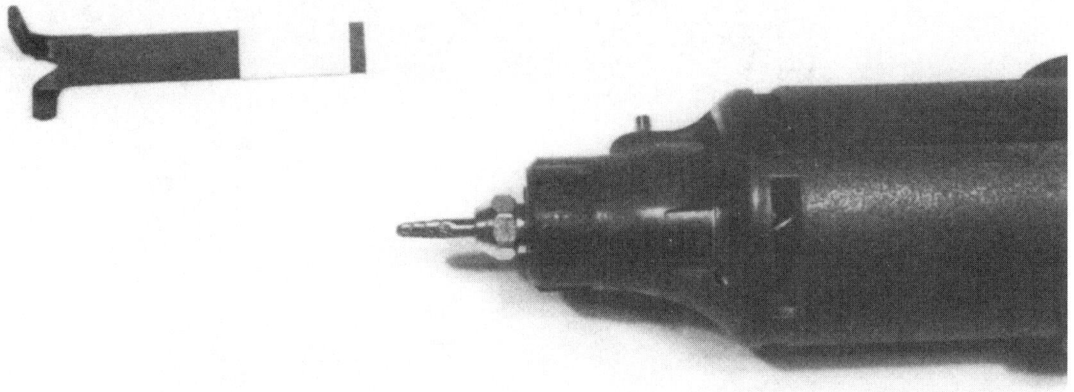

The original lever can be modifed by adding a third detent with a Dremel tool equipped with a carbide, ball end mill. The position for the detent is measured off and masking tape is used to locate the area to be milled.

An acceptable job of milling can be had by free hand use of the Dremel tool. Make several light cuts rather than big ones. Finish the bare metal with cold blue.

The outside of the trigger housing gets our attention next. The original semi-auto carbine had only two positions: "S" for safe and "F" for fire.

The stamping of the trigger housing can be done by free hand. However, a fixture such as shown above is invaluable for a professional and speedy job.

Front view of stamping tool showing the "A" stamp in place. You will note that, by using 8 set screws and lock nuts, the stamp can be finely adjusted for alignment before locking down.

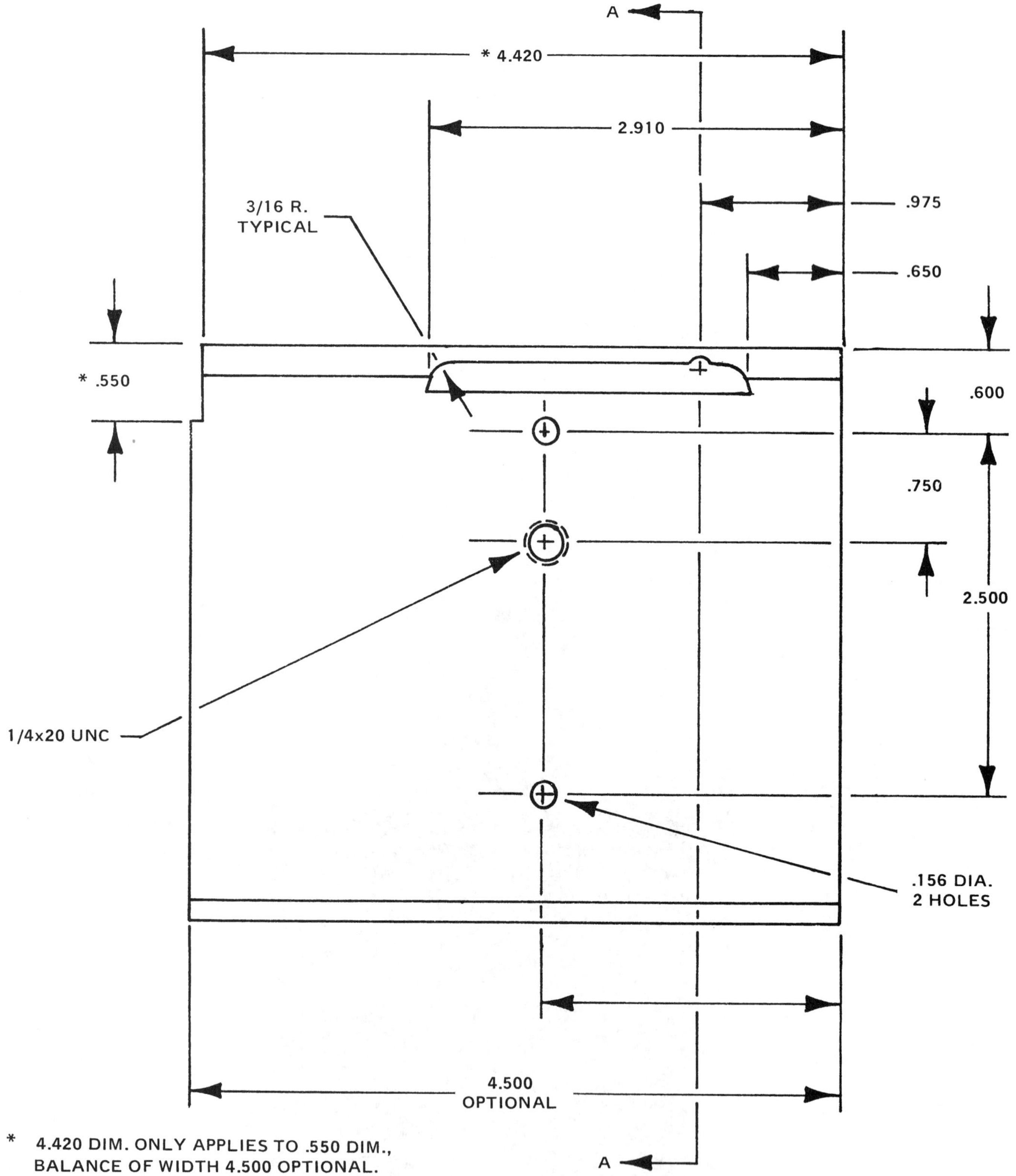

* 4.420 DIM. ONLY APPLIES TO .550 DIM., BALANCE OF WIDTH 4.500 OPTIONAL.

3/16 x 1-3/4 x 4" STEEL CHANNEL

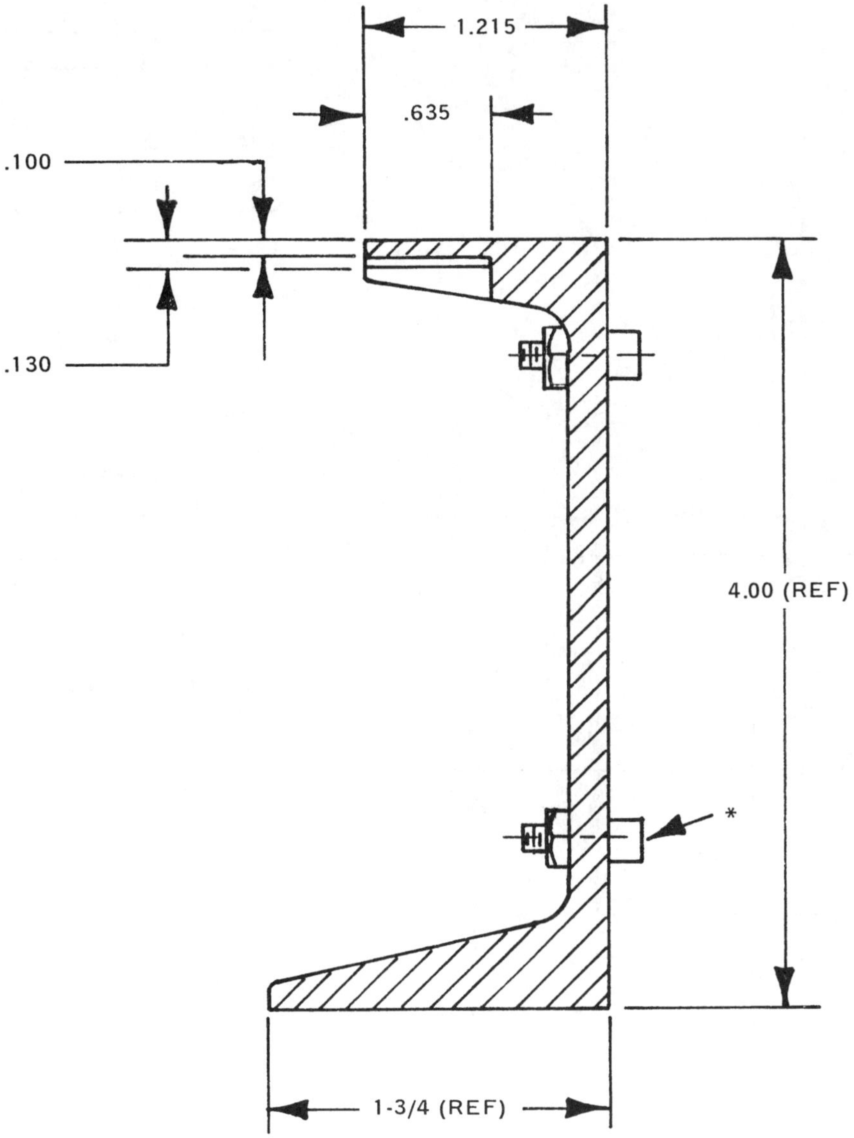

SECTION A-A

* 6-32x1/2 ALLEN HEAD CAP SCREW.
MACHINE HEAD TO .218 DIA. 2 REQ'D.
6-32 HEX NUT.

3/16 x 1-3/4 x 4" STEEL CHANNEL

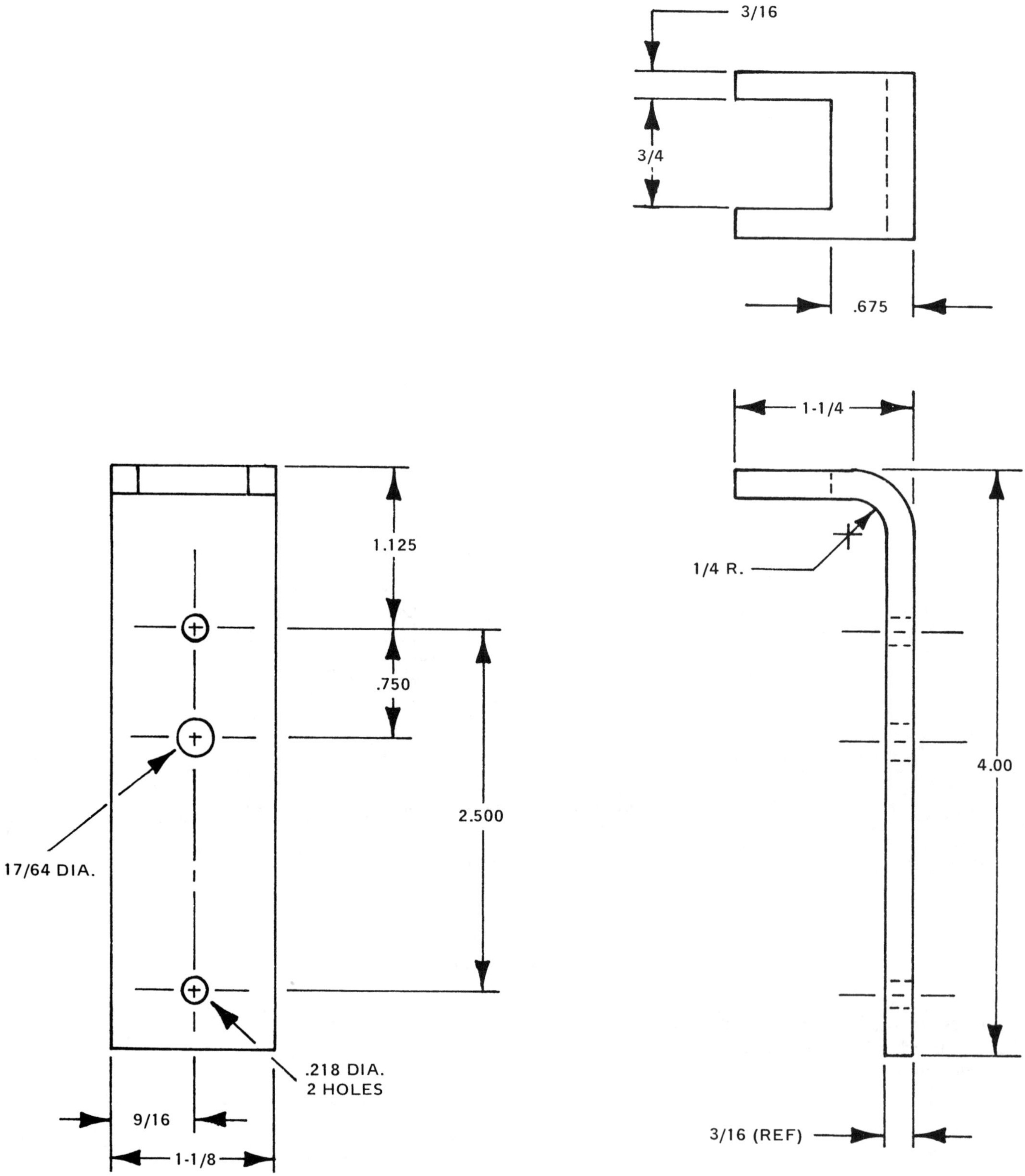

3/16 STEEL SUPPORT BRACKET

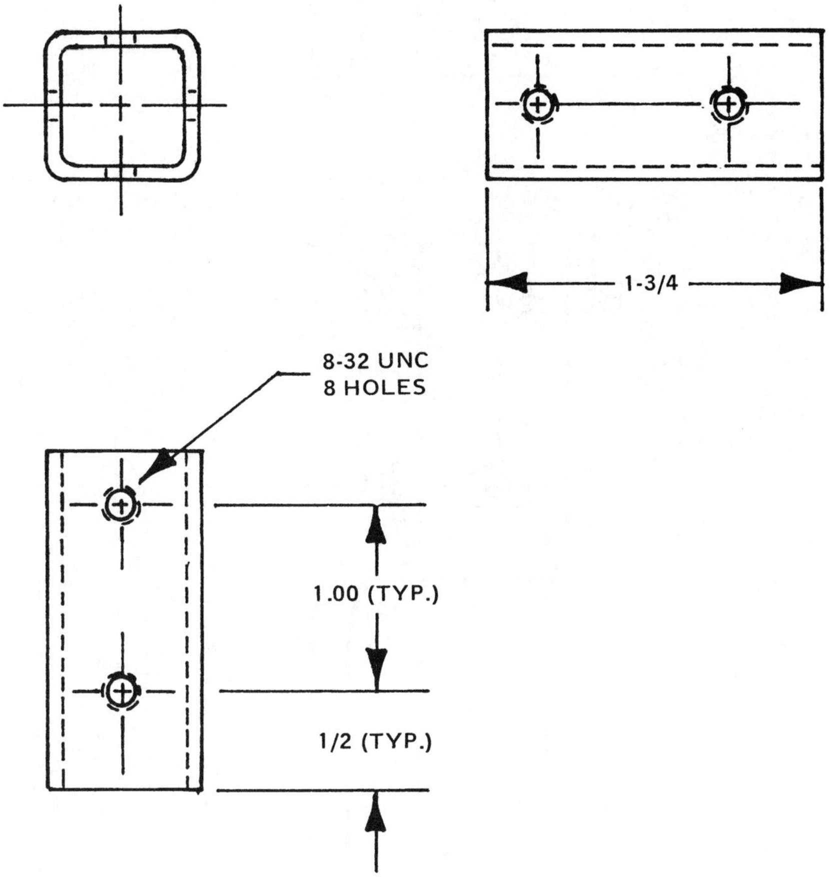

3/4x3/4x1/16 SQUARE TUBE (STEEL)

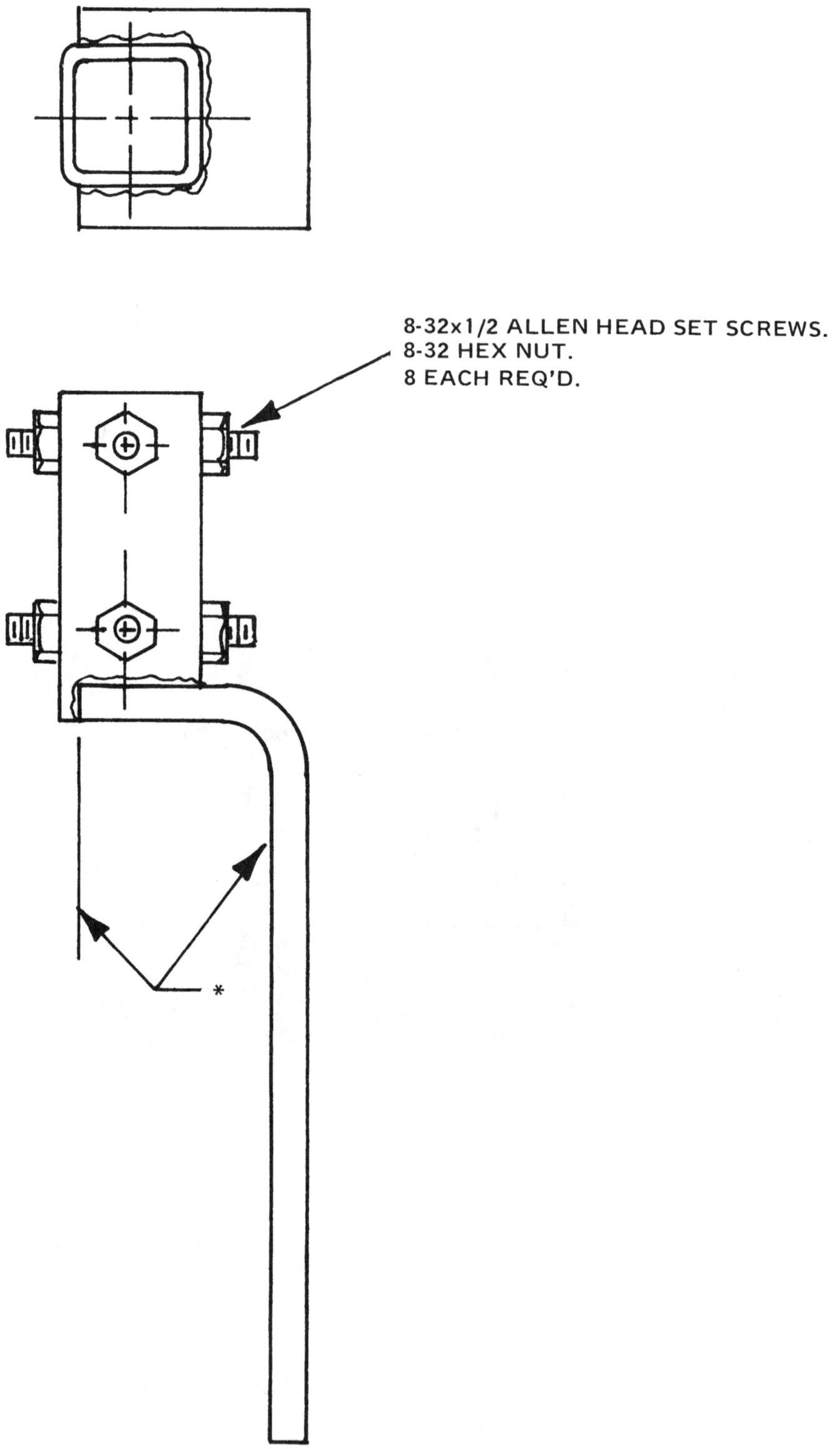

* HOLD SURFACES PARALLEL & WELD

STAMP GUIDE ASSEMBLY

The stamp holder can be quickly removed, using 2 guide pins and 1 capscrew.

With the guide removed, our fixture becomes an anvil for overstamping the "F" into an "R".

The stamping now complete, we see how the "F" was transformed into an "R" and the "A" stamped in just the rignt place. A small cold chisel is used to stamp the vertical line under the "A".

The stamping fixture accurately locates the trigger housing for perfect alignment of the "A".

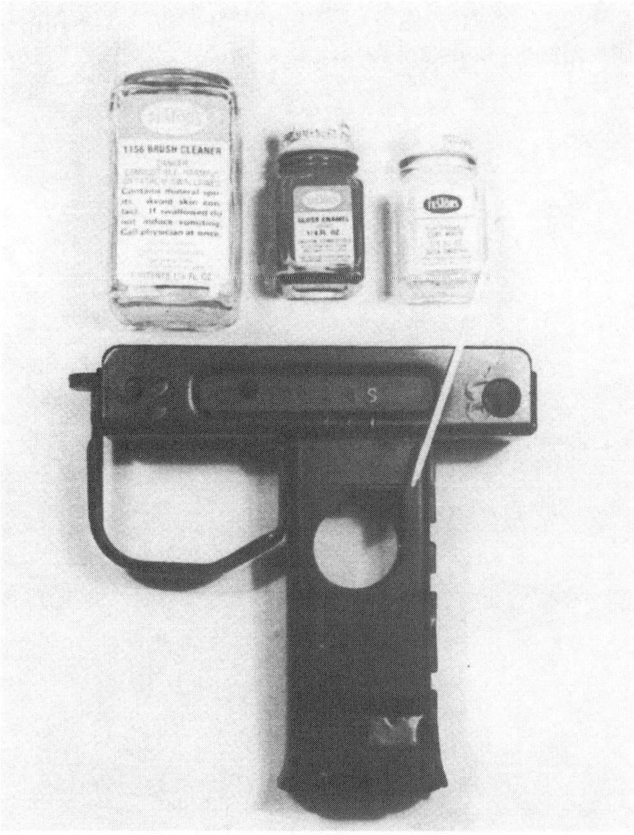

Quick drying model enamel is used to fill the letters. A sharp, round toothpick is used to fill the "A" and the "R" and their respective vertical marks with red enamel. Thinner can be used to remove any excess.

outside of the bar. The small amount of metal remaining on the inside will break through when twisted.

An adjustable wrench is used to twist the front half of the bar up and down until it shears off flush with the side of the receiver. We can now grab the rear half and apply the same up and down twisting to it. Because of the limitations presented by the rear half of the receiver, our up and down motions will be shorter. Also, because the end of the bar we're now grabbing is cantilevered forward of the tab we're trying to shear, it will start bending before the shearing is complete. In this case a hammer and punch can be used to punch the end further down than the wrench will allow. After punching down, pry it up with a large, flat-bladed screwdriver. A couple of "ups and downs" and it should shear off flush with the receiver inside wall.

In most cases the rear tab will require very little cleanup which can be done with a small flat file. The front tab area can be cleaned up, if necessary, with a Dremel tool grinder or flat file. If you plan to Parkerize the entire weapon then these exposed areas will be taken care of. If not, then cold bluing will protect the bare metal.

Since our replacement bolt cocks further to the rear we must increase the length of the cocking knob slot by 5/8" to the rear. After removing the knob assembly from the cover plate, mark the area to be removed. Using a drill bit that's smaller than the existing slot, drill out most of the area to be removed, and square up with a flat file. The bare metal should then be cold blued for protection as before. In re-installing the cocking knob assembly, replace the slot cover plate with a new one, devoid of wording.

Until now, we've been working on the inside of the weapon where, if less than perfect, our workmanship is not readily seen. In restamping the selector switch markings you have only one chance to do it right so care is in order. Using a 1/8" set of steel stamps, the "F" is overstruck with an "R". If not using a stamping fixture as illustrated on pages 25-32, it is best to wedge a piece of metal of sufficient width between the inside of the receiver wall and the magazine housing to help prevent denting the wall inward when stamping.

Stamping the "A" is easy if a fixture is used. If not, then care must be exercised to locate the letter in the proper place. Masking tape can be used to expose the area to be stamped. If overstamping the "A" should be necessary, use care when aligning the stamp. Use a small cold chisel to indent the vertical mark under the "A".

Using a sharp toothpick, fill the new letters and marks with red enamel as pictured. If properly stamped and filled, your Uzi will look "factory original".

When re-assembling the trigger mechanism the original select lever can be used, although, if of later production, it will not have a detent in the auto position. Replacement levers are available. However, they are devoid of the safety tab. It is the author's recommendation that a detent be added to the original lever as illustrated herein. This can be done "free hand" with a Dremel tool equipped with a 1/8" ball end mill.

Final Assembly

Before any reassembly is begun, thoroughly clean any grinding residue from both the trigger housing and receiver.

The replacement bolt will have to have an extractor installed in it. Although the original from the semi-auto bolt can be used, it is recommended that you replace it with one of the available wide extractors. You will note from the photograph on page 36 that they have a noticeably larger contact area and should prove more reliable.

Reassemble the trigger housing and install it onto the receiver. Install a barrel, either the original 16" one or a new "shorty". Place the bolt, without the recoil spring, in the receiver and snap the cover plate in place. Place the selector on "A", hold the trigger down, and check for free travel of the bolt by tilting the receiver up and down. Make sure the face of the bolt is contacting the chamber of the barrel. No part of the bolt face should contact the feed ramp when closing. The author has experienced bolts that did, resulting in improper functioning.

Now reinstall the bolt with the recoil spring assembly. Try the cocking knob to insure the bolt being fully engaged on the sear. Try all select lever positions for proper function, especially safety. If all is well, the stock, either folding or wooden, can be installed and your new machine gun is ready to test fire.

Uzis converted by the author have digested most ammo that was fed them. However, some PMC 9mm cases were so thin near the base that a severe bulge was produced. In most blowback SMG's there's still a lot of chamber pressure existent when the rearward travel of the bolt begins. Case wall thickness, of properly made ammo, increases near the base and is not affected by this pressure during extraction. PMC ammo of later manufacture has not been used and, whether or not PMC have cleaned up their act, cannot be verified at this time.

So be forewarned and proceed, one round at a time, with questionable ammunition. A blown case could result in a bullet stuck midway in the barrel. If not cleared, this could produce a blown or bulged barrel – if not greater damage!

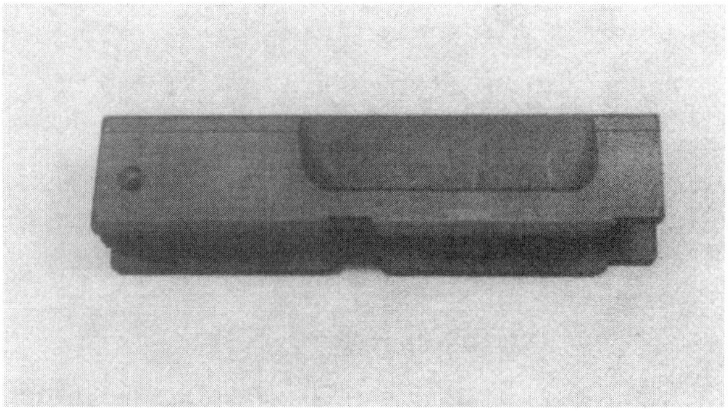

Shown here is a Group Industries full stroke bolt.

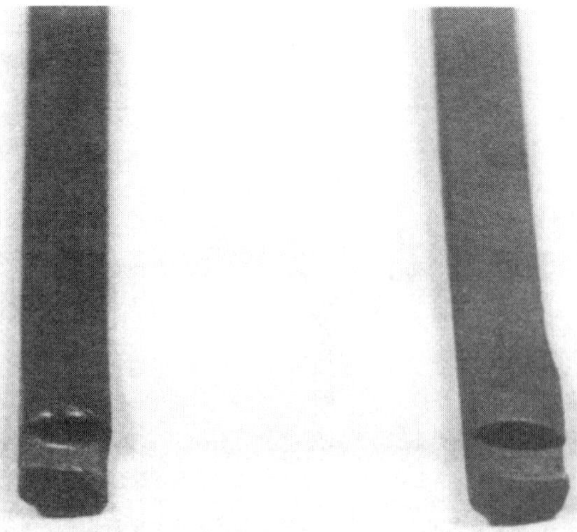

It is recommended to install one of the available machine gun extractors at this stage. Note the larger gripping area of these (right) when compared to the original Uzi semi-auto extractor (left).

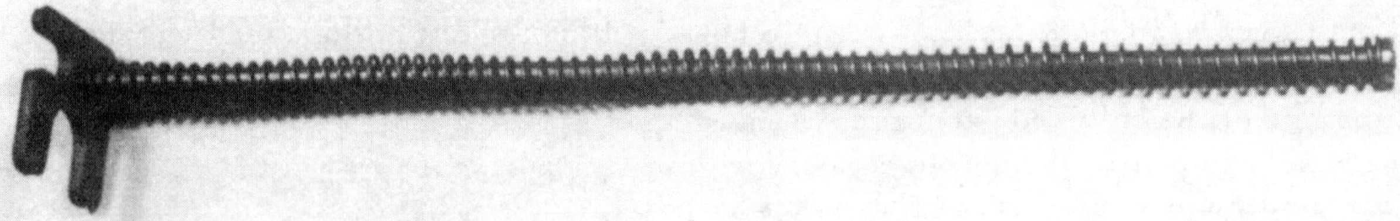

A new recoil spring/buffer assembly, again from Group Industries, is required.

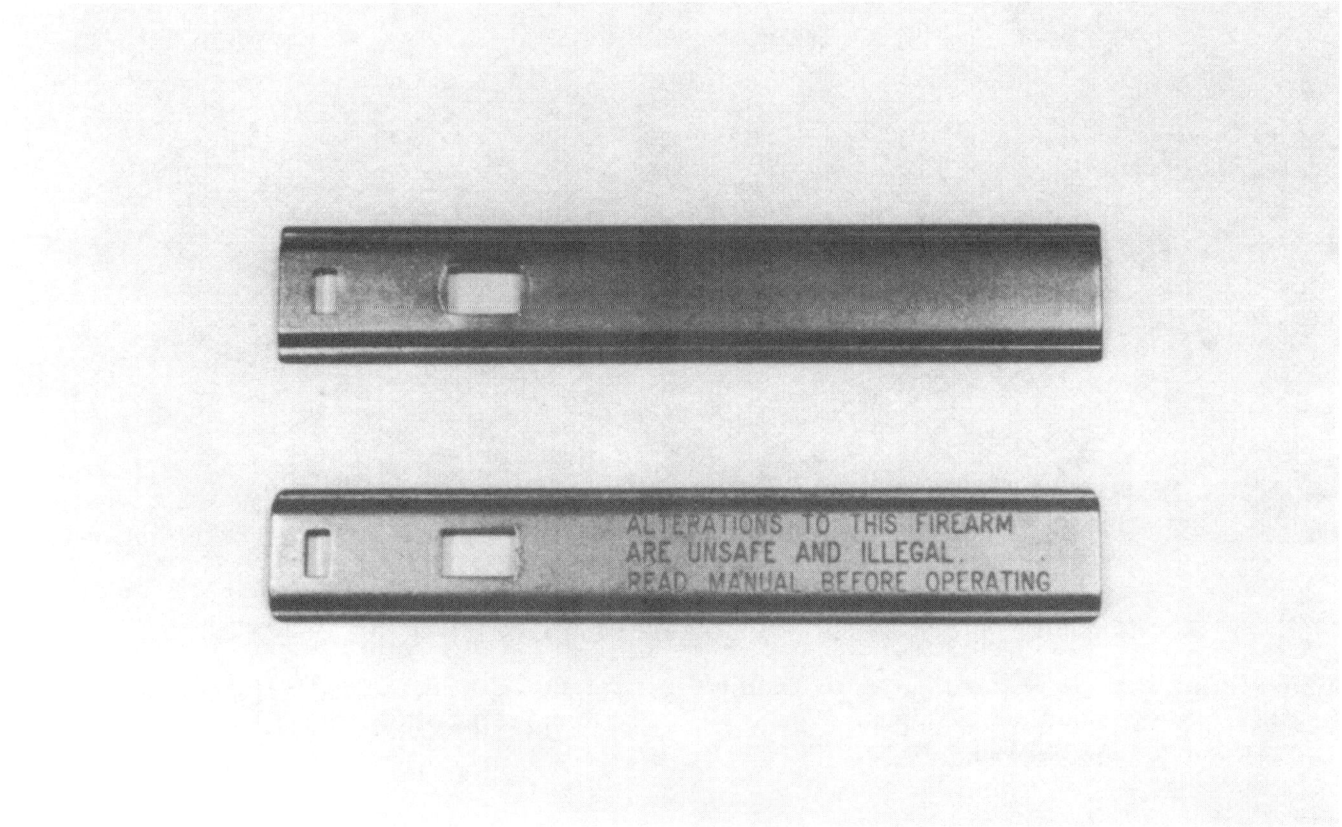

Although the change is only cosmetic, a replacement cocking slot cover plate (top) should be installed. Available from Group Industries, it is devoid of the wording that appears on the original (bottom).

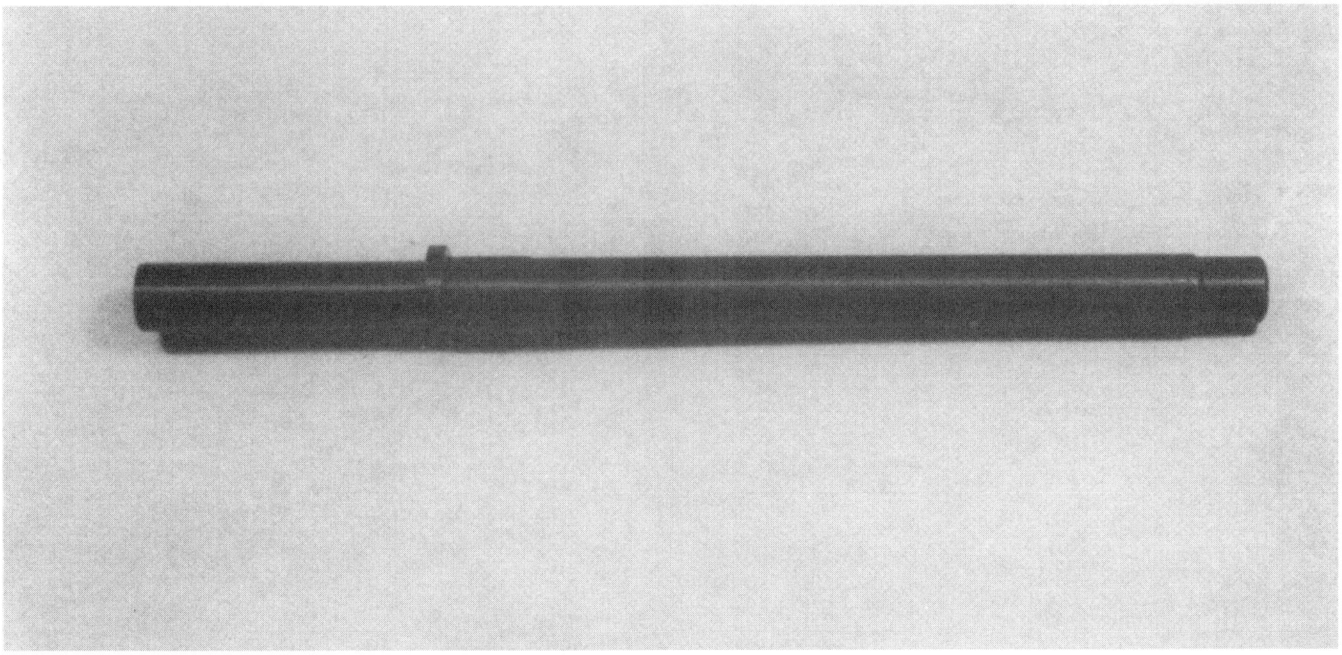

Replacement barrels are plentiful. The one used here is from Tony Scherer, Gilberts, IL.

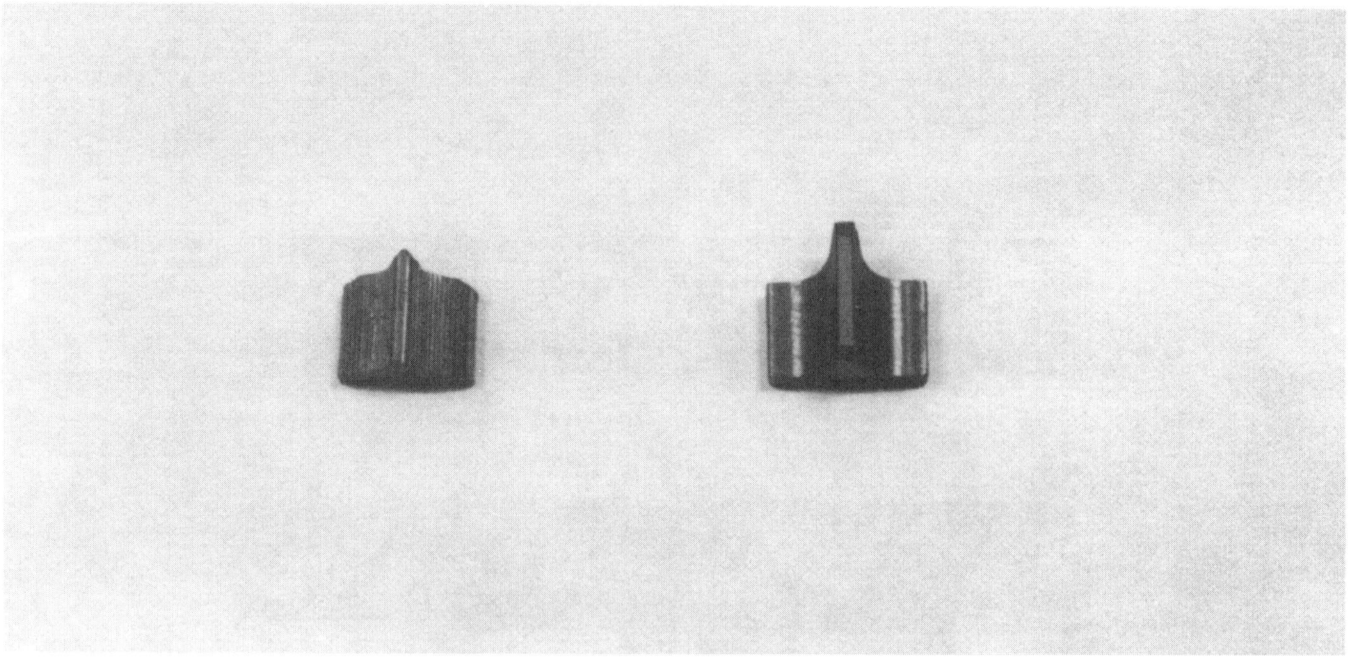

Although not a necessity, a new selector knob is available. Made of anodized aluminum, it features a larger contact surface that makes for easier operation than the original. These can occasionally be found at gun shows.

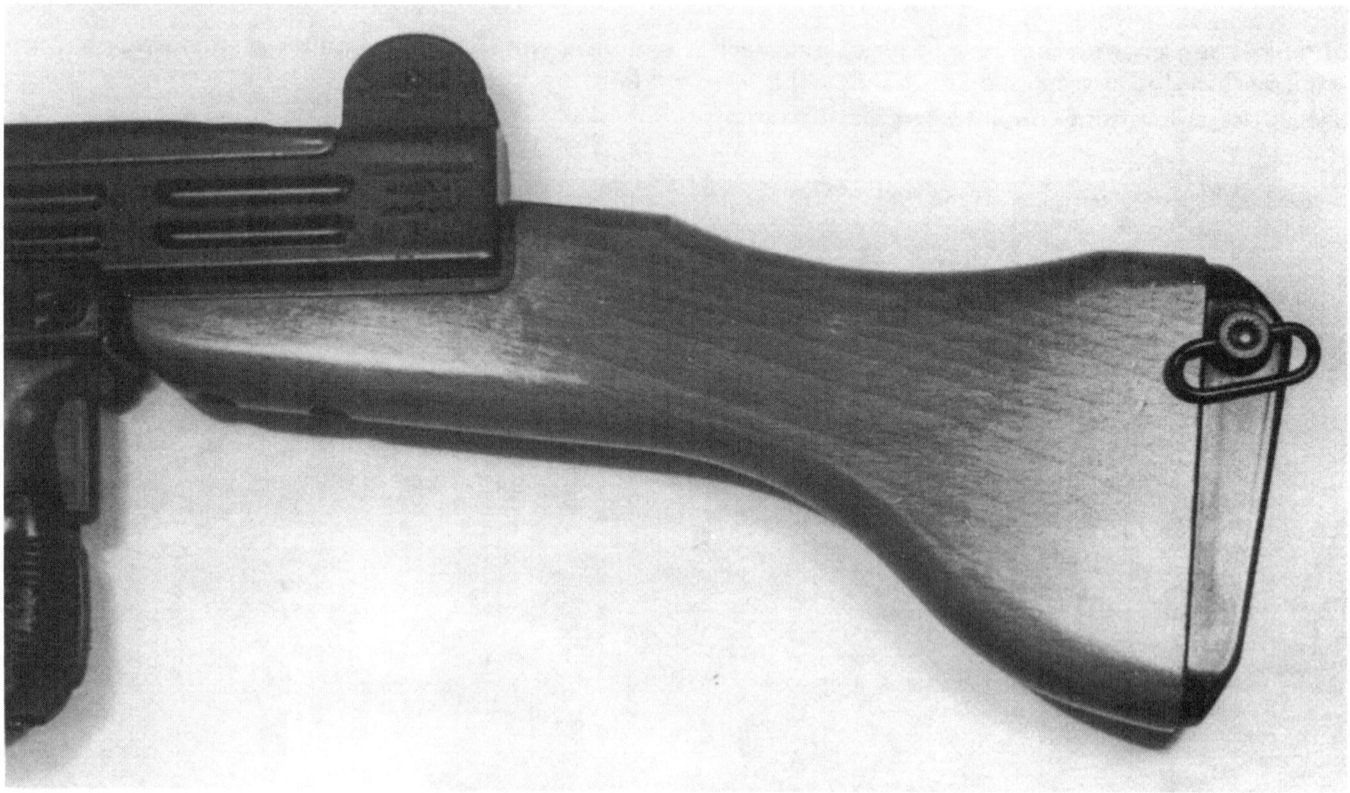

Unless compact storage is a requirement, Uzi's accessory wooden stock is a great improvement over the collapsible out-of-the-box model.

The complete selective fire Uzi ready to fire.

Our new Uzi, with wooden buttstock installed, being test fired. By using the full stroke, heavier bolt, it was possible to squeeze off one round at a time while set on automatic.

Notes

Notes

Notes

Notes

Other Books Available From Desert Publications

081	Crossbows/ From 35 Years With the Weapon	$11.95
206	CIA Field Expedient Preparation of Black Powder	$11.95
221	Evaluation of Improvised Shaped Charges	$11.95
411	Homeopathic First Aid	$11.95
432	Cold Weather Survival	$11.95
435	Homestead Carpentry	$14.95
444	Leadership Hanbook of Small Unit Ops	$9.98
454	Survival Childbirth	$13.95
C-002	How To Open A Swiss Bank Account	$9.95
C-011	Defending Your Retreat	$9.95
C-020	Methods of Long Term Storage	$9.95
C-023	Federal Firearms Laws	$11.95
C-024	Select Fire Uzi Modification Manual	$13.95
C-025	How to Build Silencers	$6.95
C-027	Firearms Silencers Volume 1	$11.95
C-029	M1 Carbine Arsenal History	$7.95
C-032	The Silencer Cookbook	$9.95
C-038	How To Build A Beer Can Morter	$4.95
C-040	Criminal Use of False ID	$14.95
C-045	Improvised Weapons of American Underground	$11.95
C-048	Catalog of Military Suppliers	$18.95
C-056	Navy Seal Manual	$29.95
C-065	Poor Man's James Bond Vol 1	$34.95
C-113	Submachine Gun Designers Handbook	$19.95
C-114	AR-15, M16 and M16A1 5.56mm Rifles	$13.95
C-117	Ruger Carbine Cookbook	$11.95
C-120	Shootout II	$14.95
C-122	AK-47 Assault Rifle	$15.95
C-123	Combat Loads for Sniper Rifles	$16.95
C-125	Browning Hi-Power Pistols	$13.95
C-126	UZI Submachine Gun	$13.95
C-127	FullAuto Vol 1 Ar-15 Modification Manual	$11.95
C-128	Full Auto Vol 2 Uzi Mod Manual	$11.95
C-130	Full Auto Vol 4 Thompson SMG	$11.95
C-131	FullAuto Vol 5 M1 Carbine to M2	$11.95
C-135	M-14 Rifle, The	$12.95
C-136	Fighting Garand Manual	$13.95
C-137	Ranger Training & Operations	$19.95
C-138	Emergency War Surgery	$29.95
C-140	Emergency Medical Care/Disaster	$14.95
C-143	Survival Medicine	$15.95
C-150	Improvised Munitions Black Book Vol 1	$17.95
C-151	Improvised Munitions Black Book Vol 2	$17.95
C-152	Impro. Munitions Black Book Vol 3	$29.95
C-155	Fighting Back on the Job	$14.95
C-160	Firearm Silencers Vol 2	$24.95
C-162	Colt .45 Auto Pistol Manual	$11.95
C-163	Survival Evasion & Escape	$19.95
C-165	USMC Sniping	$18.95
C-175	Beat the Box	$7.95
C-197	How to Make Disposable Silencers Vol. 1	$18.95
C-198	Expedient Hand Grenades	$21.95
C-199	U. S. Army Counterterrorism Training Manual	$19.95
C-203	Explosives and Propellants	$14.95
C-204	Improvised Munitions/Ammonium Nitrate	$13.95
C-205	Two Component High Explosives Mixtures	$14.95
C-207	CIA Improvised Sabotage Devices	$14.95
C-210	Guerilla Warfare	$14.95
C-211	FN-FAL Auto Rifles	$21.95
C-218	Professional Homemade Cherry Bomb	$15.95
C-219	Improvised Shaped Charges	$14.95
C-220	Improvised Explosives, Use In Detonation Devices	$14.95
C-227	Springfield '03 Rifle Manual	$15.95
C-229	Lock Picking Simplified	$9.95
C-230	How to Fit Keys by Impressioning	$9.95
C-231	Lockout -Techniques of Forced Entr	$13.95
C-235	How to Open Handcuffs Without Keys	$11.95
C-244	Camouflage	$12.95
C-274	M14 and M14A1Rifles and Rifle Marksmanship	$23.95
C-284	Thompson Submachine Guns	$19.95
C-289	M1 Carbine Owners Manual	$13.95
C-290	Clandestine Ops Man/Central America	$11.95
C-324	How to Make Disposable Silencers Vol. 2	$18.95
C-333	USMC AR-15/M-16 A2 Manual	$21.95
C-337	Sten MKII SMG Construction Manual	$24.95
C-356	Map Reading and Land Navigation	$19.95
C-365	Survival Gunsmithing	$14.95
C-372	Poor Man's James Bond Vol 2	$34.95
C-386	Self-Defense Requires No Apology	$13.95
C-414	Select Fire 10/22	$15.95
C-415	Construction Secret Hiding Places	$15.95
C-416	Brown's A!cohol Motor Fuel Cookbook	$24.95
C-417	Canteen Cup Cookery	$9.95
C-418	Improvised Batteries/Detonation Devices	$14.95
C-421	Walther P-38 Pistol Manual	$13.95
C-422	P-08 Parabellum Luger Auto Pistol	$14.95
C-423	Sten Submachine Gun, The	$4.98
C-427	How to Train a Guard Dog	$16.95
C-429	The Anarchist Handbook Vol. 1	$14.95
C-430	Beretta - 9MM M9	$15.95
C-432	Mercenary Operations Manual	$5.95
C-439	Improvised Lock Picks	$13.95
C-455	Modern Day Ninjutsu	$17.95
C-509	Survival Guns	$24.95
C-512	FullAuto Vol 8 M14A1 & Mini 14	$11.95
C-518	L.A.W. Rocket System	$14.95
C-528	With British Snipers, To the Reich	$29.95
C-539	Dental Specialist	$49.95
C-558	Food. Fur and Survival	$17.95
C-561	The Squeaky Wheel	$14.95
C-562	Survival Chemist Book	$17.95
C-564	Medical Specialist	$49.95
C-571	U. S. Marine Corps Scout/Sniper Training Manual	$27.95
C-610	Aunt Bessie's Wood Stove Cookbook	$9.95
C-622	Defensive Shotgun	$14.95
C-625	MAC-10 Cookbook	$11.95
C-627	Brown's Lawsuit Cookbook	$26.95
C-628	Can You Survive	$16.95
C-633	Hand to Hand Combat	$11.95
C-634	USMC Hand to Hand Combat	$9.95
C-635	Hand to Hand Combat by D'Eliscue	$7.95
C-636	Prisons Bloody Iron	$13.95
C-639	Police Karate	$17.95
C-646	Science of Revolutionary Warfare	$16.95
C-668	Survival Shooting for Women	$7.48
C-680	Infantry Scouting, Patrol. & Sniping	$16.95
C-694	CIA Field Expedient Incendiary Manual	$17.95
C-761	How to Lose Your X-Wife Forever	$24.95
C-763	Vigilante Handbook	$17.95
C-775	Alcohol Distillers Handbook	$24.95
C-776	How to Build a Junkyard Still	$15.95
C-777	Engineer Explosives of WWI	$7.48
C-794	US Marine Corps Essential Subjects	$16.95
C-815	US Marine Bayonet Training	$9.95
C-818	The Poisoner's Handbook	$29.95
C-829	The Anachist Handbook Vol. 2	$14.95
C-889	Napoleon's Maxims of War	$11.95
C-890	Company Officers HB of Ger. Army	$9.48
C-891	German Infantry Weapons Vol 1	$18.95
C-894	Shotguns, eight Gov. Types	$17.95
C-896	Op. Man. 7.62mm M24 Sniper Weapon	$11.95
C-899	US Army Bayonet Training	$4.48
C-9002	German MG-34 Machinegun Manual	$19.95
C-9012	Firearm Silencers Vol 3	$16.95
C-9013	HK Assault Rifle Systems	$27.95
C-9047	SKS Type of Carbines, The	$17.95
C-9048	Rough Riders, The	$24.95
C-9052	Private Weaponeer, The	$14.95
C-9057	Keys To Understanding Tubular Locks	$11.95
C-9060	The Anarchist Handbook Vol. 3	$14.95
C-9081	Dirty Fighting	$11.95
C-9082	Lasers & Night Vision Devices	$29.95
C-9083	Water Survival Training	$8.95
C-9084	Assorted Nasties	$29.95
C-9085	Ruger P-85 Family of Handguns	$19.95
C-9101	Guide to Germ Warfare	$14.95
C-9102	CIA Field Expedient Method for Explosives Preparation	$14.95
C-9119	Surviving Global Slavery	$16.95
C-9120	Military Knife Fighting	$14.95
C-9127	H&R Reising Submachine Gun Manual	$16.95
C-9131	Live to Spend It	$29.95
C-9134	Military Ground Rappelling Techniques	$13.95
C-9135	Smith & Wesson Autos	$19.95
C-9137	Caching Techniques of U.S. Army Special Forces	$14.95
C-9138	USMC Battle Skills Training Manual	$49.95
C-9139	Survival Armory	$27.95
C-9156	Concealed Carry Made Easy	$17.95
C-9164	The L'il M-1, The .30 Cal. M-1 Carbine	$17.95
C-9170	Urban Combat	$27.95
C-9178	Apocalypse Tomorrow	$17.95
C-9198	MP40 Machinegun Manual	$16.95
C-9199	Clear Your Record & Own a Gun	$24.95
C-9200	Sig's Handguns Manual	$19.95
C-9210	Sniper Training	$24.95
C-9212	Poor Man's Sniper Rifle	$18.95
C-9221	The Official Makarov Pistol Manual	$14.95
C-9224	The Butane Lighter Hand Grenade	$12.95
C-9228	Unarmed Against the Knife	$14.95
C-9229	Black Book of Booby Traps	$29.95
C-9239	Militia Battle Manual	$22.95
C-9247	Black Book of Revenge	$15.95
C-9253	Bankruptcy, Credit and You	$18.95
C-9255	The Sicilian Blade	$15.95
C-9259	How to Build Practical Firearm Silencers	$16.95
C-9262	Glock's Handguns	$19.95
C-9263	Heckler and Koch's Handguns	$19.95
C-9264	The Poor Man's Ray Gun	$11.95
C-9265	The Poor Man's R. P. G.	$22.95
C-9279	Credit Improvement & Protection Handbook	$19.95
C-9281	Build Your Own AR-15	$19.95
C-9282	How to Live Safely in a Dangerous World	$19.95
C-9293	Bllack Book of Arson	$24.95
C-9322	How to Build Flash/Stun Grenades	$19.95
C-9342	Entrapment	$19.95
C-9359	The FN-FAL Rifle Et Al.	$21.95
C-9367	The Mental Edge, Revised	$17.95
C-9373	Survival Bible by Duncan Long	$59.95
C-9405	The Internet Predator	$17.95
C-9406	The Criminal Defendants Bible	$49.95
C-9430	Protect Your Assets	$24.95
C-9432	How To Build Military Grade Suppressors	$21.95
C-9441	Mossberg Shotguns	$24.95
C-9478	Poor Man's Primer Manual, The	$24.95
C-9546	Agents HB of Black Bag Ops.	$14.95
C-9556	Improvised Rocket Motors	$7.48

PRICES SUBJECT TO CHANGE WITHOUT NOTICE

800-852-4445

info@deltapress.com

Desert Publications

505 East 5th St.

Dept. - BK-004

El Dorado, AR 71730 USA

Shipping & handling
1 item $9.95 - 2 or more $11.95